AF847314

Springer Series in Biomaterials Science and Engineering

Volume 20

Series Editor

Min Wang, Department of Mechanical Engineering, The University of Hong Kong, Pokfulam Road, Hong Kong

The Springer Series in Biomaterials Science and Engineering addresses the manufacture, structure and properties, and applications of materials that are in contact with biological systems, temporarily or permanently. It deals with many aspects of modern biomaterials, from basic science to clinical applications, as well as host responses. It covers the whole spectrum of biomaterials–polymers, metals, glasses and ceramics, and composites/hybrids–and includes both biological materials (collagen, polysaccharides, biological apatites, etc.) and synthetic materials. The materials can be in different forms: single crystals, polycrystalline materials, particles, fibers/wires, coatings, non-porous materials, porous scaffolds, etc. New and developing areas of biomaterials, such as nano-biomaterials and diagnostic and therapeutic nanodevices, are also focuses on in this series. Advanced analytical techniques that are applicable in R&D and theoretical methods and analyses for biomaterials are also important topics. Frontiers in nanomedicine, regenerative medicine, and other rapidly advancing areas calling for great explorations are highly relevant. The Springer Series in Biomaterials Science and Engineering aims to provide critical reviews of important subjects in the field, publish new discoveries and significant progresses that have been made in both biomaterials development and the advancement of principles, theories, and designs, and report cutting-edge research and relevant technologies. The individual volumes in the series are thematic. The goal of each volume is to give readers a comprehensive overview of an area where new knowledge has been gained and insights made. Significant topics in the area are dealt with in good depth and future directions are predicted on the basis of current developments. As a collection, the series provides authoritative works to a wide audience in academia, the research community, and industry.

This book series is indexed by the EI Compendex and Scopus databases.

If you are interested in publishing your book in the series, please contact Dr. Mengchu Huang (Email: mengchu.huang@springer.com).

Jianxi Xiao

Collagen Mimetic Peptides and Their Biophysical Characterization

Springer

Jianxi Xiao
College of Chemistry and Chemical Engineering
Lanzhou University
Lanzhou, Gansu, China

ISSN 2195-0644 ISSN 2195-0652 (electronic)
Springer Series in Biomaterials Science and Engineering
ISBN 978-981-10-1940-1 ISBN 978-981-10-1941-8 (eBook)
https://doi.org/10.1007/978-981-10-1941-8

© Springer Nature Singapore Pte Ltd. 2024

This work is subject to copyright. All rights are solely and exclusively licensed by the Publisher, whether the whole or part of the material is concerned, specifically the rights of translation, reprinting, reuse of illustrations, recitation, broadcasting, reproduction on microfilms or in any other physical way, and transmission or information storage and retrieval, electronic adaptation, computer software, or by similar or dissimilar methodology now known or hereafter developed.
The use of general descriptive names, registered names, trademarks, service marks, etc. in this publication does not imply, even in the absence of a specific statement, that such names are exempt from the relevant protective laws and regulations and therefore free for general use.
The publisher, the authors and the editors are safe to assume that the advice and information in this book are believed to be true and accurate at the date of publication. Neither the publisher nor the authors or the editors give a warranty, expressed or implied, with respect to the material contained herein or for any errors or omissions that may have been made. The publisher remains neutral with regard to jurisdictional claims in published maps and institutional affiliations.

This Springer imprint is published by the registered company Springer Nature Singapore Pte Ltd.
The registered company address is: 152 Beach Road, #21-01/04 Gateway East, Singapore 189721, Singapore

If disposing of this product, please recycle the paper.

Preface

In the intricate realm of biomolecules, collagen commands unparalleled significance as the most abundant protein within the human body, owing to its exceptional structural properties and versatile biological functions that have captured the curiosity of scientists across generations. Collagen's distinctive hierarchical self-assembly leads to the creation of vital molecular scaffolds, which are the bedrock for maintaining the structural integrity and mechanical resilience of connective tissues. Concurrently, collagen plays indispensable roles in various physiological processes, encompassing tissue regeneration, cell proliferation, and wound healing, while its abnormalities are implicated in a wide array of diseases, notably cancer and arthritis. Nonetheless, exploring collagen's intricacies through biophysical techniques presents formidable challenges due to its substantial size, repetitive sequence, and complex hierarchical structure. Collagen mimetic peptides provide a more accessible platform for biophysical characterization and represent an intriguing opportunity to delve into the structure, function, and pathology of collagen.

The book titled *Collagen Mimetic Peptides and Their Biophysical Characterization* embarks on an enlightening journey through the molecular landscape of collagen mimetic peptides, illuminating their significance and potential. This journey commences with a foundational exploration of collagen's unique attributes, spanning its classification, functions, hierarchical structures, and sequential patterns. The book offers an in-depth examination of biophysical techniques for unraveling the structural intricacies of collagen mimetic peptides, encompassing Circular Dichroism (CD), X-ray crystallography, Nuclear Magnetic Resonance (NMR), Fluorescence, and Raman spectroscopy. Additionally, it thoroughly covers various facets of collagen mimetic peptides, ranging from synthetic strategies and the targeting of pathological collagen to the construction of homotrimeric, heterotrimeric, and self-assembled peptide models.

This book explores the profound implications of collagen mimetic peptides across various scientific domains, including protein science, biomaterials, bioanalysis, and beyond. It caters to a diverse readership, from seasoned researchers seeking an in-depth understanding of collagen and its mimics to undergraduate and graduate students eager to delve into this fundamental protein's intricacies.

Throughout the arduous process of composing this book, I have been fortunate to receive invaluable support and assistance from esteemed professors, doctors, and graduate students affiliated with the Joint Collagen Research Center at Lanzhou University. I extend my deepest gratitude to each of them.

Recognizing the constraints of my expertise and professional background, I earnestly invite and value constructive criticism from readers. The collaborative pursuit of knowledge is an ever-evolving process, and your insights are integral to its progression.

Lanzhou, China Jianxi Xiao

Contents

About the Author

Dr. Jianxi Xiao received his Bachelor's degree from Wuhan University in 2003 and completed his Ph.D. at Rutgers University-New Brunswick, New Jersey in 2009. He pursued postdoctoral research at Rutgers University and subsequently attained a full professorship at Lanzhou University's College of Chemistry and Chemical Engineering in 2012. Dr. Xiao holds the distinguished title of Cuiying Scholar at Lanzhou University and is honored with the prestigious "Feitian Scholar" professorship in Gansu Province. In addition, he serves as the Director of Medical Collagen Engineering Research Center in Gansu Province and as the Director of Joint Collagen Research Center at Lanzhou University. His research interests encompass the development and characterization of innovative collagen mimic peptides, collagen-targeting peptide probes, and collagen-based biomaterials. Dr. Xiao's extensive work has significantly advanced our comprehension of collagen's structure, function, and pathology, leading to the publication of over 70 academic papers in esteemed journals.

Chapter 1
The Collagen Suprafamily

Abstract Collagen, the predominant protein in mammals, serves as an essential structural component in various connective tissues such as skin, bones, tendons and ligaments. It imparts mechanical strength and structural integrity, crucial for supporting and protecting softer tissues and internal organs. Displaying significant heterogeneity, collagen encompasses 29 identified types, which form both fibrillar and non-fibrillar structures. Fibrillar collagens, such as types I, II, and III, are vital for connective tissue stability, while non-fibrillar collagens, including types IX and XVIII, fulfill specialized roles in anchoring and network formation. Collagen interacts with diverse receptors, including integrins, discoidin domain receptors (DDRs), glycoprotein VI (GPVI), and leukocyte-associated IG-like receptor-1 (LAIR-1), which are integral in regulating essential physiological processes such as cell adhesion, migration, and tissue remodeling. Mutations in collagen genes are associated with a range of disorders, including osteogenesis imperfecta (OI), Type II collagenopathies, Ehlers-Danlos Syndrome type IV, and Alport syndrome. Therefore, deepening our comprehension of collagen's structural diversity and its biological interactions is imperative for elucidating its multifaceted roles and implications in pathology.

Keywords Collagen · Fibrillar · Non-fibrillar · Biological interactions · Disorders

1.1 Introduction to Collagen

Collagen is the most abundant protein in mammals, making up approximately 25–35% of the whole-body protein content. As the major structural protein, collagen is widely found in various connective tissues, such as skin, bone, tendon, ligament, cartilage, cornea and blood vessels [1–3]. Collagen forms a molecular scaffold that provides mechanical strength and structural integrity to the human body, supporting and protecting the softer tissues and internal organs. The name collagen originates from the Greek word kólla, meaning "glue" [4]. It dates back to the historical use of the compound to produce glue by the process of boiling animal skins.

© Springer Nature Singapore Pte Ltd. 2024
J. Xiao, *Collagen Mimetic Peptides and Their Biophysical Characterization*,
Springer Series in Biomaterials Science and Engineering 20,
https://doi.org/10.1007/978-981-10-1941-8_1

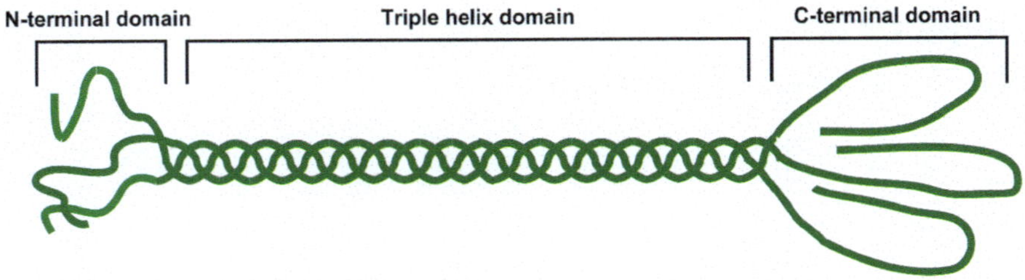

Fig. 1.1 The characteristic structure of collagen

Collagen presents a still growing family of triple-helical, extracellular matrix proteins with an extraordinary heterogeneity. Twenty-nine distinct types of collagens have been identified, including those that form fibrils with periodic bands and others classified as non-fibrillar collagens [5]. Collagen exhibits diverse structural characteristics, with some being homotrimers like fibrillar Type II and Type III collagens, while others are heterotrimers like fibrillar Type I and non-fibrillar Type IV collagens [6]. The molecular hallmark of all types of collagens is the unique triple helix configuration formed by the close packing of three polypeptide chains (Fig. 1.1). Collagen contains the characteristic triple helix domain conjugated by the non-triple helical N- and C- terminal domains [7–9].

1.2 Types of Collagens

The collagen suprafamily is characterized by an extraordinary diversity in terms of molecular composition, supramolecular structures, tissue distribution, and pathology (Table 1.1) [4, 10]. To date, 40 genes encoding vertebrate collagen polypeptide chains have been identified, resulting in the formation of 29 distinct homo- and/ or heterotrimeric molecules [11, 12]. Collagens are classified into different subfamilies based on their supramolecular assembly patterns, including fibrillar, network, and beaded filament collagens.

The distribution of various types of collagens within tissues is highly varied, with each type of collagen having a specific function and localization within the body. For example, Type I collagen is widely found in skin, bone, tendon, and ligament, providing tensile strength to these tissues [13]. On the other hand, Type II collagen is predominantly found in cartilage and, together with proteoglycans, provides compressive properties to joints [14]. Type III collagen is typically found in relatively elastic tissues such as embryonic skin, blood vessels, and various internal organs [15].

Table 1.1 The collagen suprafamily [4, 10]

Type	Composition	Subfamily	Distribution	Pathology
I	$[\alpha_1(I)]_2\alpha_2(I)$	Fibrillar	Skin, Bone, Tendon, Ligament, Cornea	Osteagenesis imperfecta, Ehlers-Danlos syndrome, Osteoporosis
I	$[\alpha_1(I)]_3$	Fibrillar	Tumor, Dermis, Bone	Cancer
II	$[\alpha_1(II)]_3$	Fibrillar	Cartilage, Vitreous	Chondrodysplasias, Osteoporosis
III	$[\alpha_1(III)]_3$	Fibrillar	Skin, Vessel, Intestine, Uterus	Ehlers-Danlos syndrome, Aortic aneurysm
IV	$[\alpha_1(IV)]_2\alpha_2(IV)$ $\alpha_3(IV)$ $\alpha_4(IV)$ $\alpha_5(IV)$ $[\alpha_5(IV)]_2\alpha_6(IV)$	Network	Basement membranes	Alport syndrome, Hematuria
V	$[\alpha_1(V)]_2\alpha_2(V)$ $[\alpha_1(V)]_3$ $\alpha_1(V)\alpha_2(V)\alpha_3(V)$	Fibrillar	Bone, Skin, Cornea, Placenta	Ehlers-Danlos syndrome
VI	$\alpha_1(VI)\alpha_2(VI)\alpha_3(VI)$	Beaded filament	Bone, Cartilage, Skin, Cornea, Vessel	Bethlem myopathy
VII	$[\alpha_1(VII)]_3$	Anchoring fibrils	Skin, Bladder, Oral mucosa	Epidermolysis bullosa dystrophica
VIII	$[\alpha_1(VIII)]_2\alpha_2(VIII)$ $[\alpha_1(VIII)]_3$ $[\alpha_2(VIII)]_3$	Network	Brain, Heart, Kidney, Skin, Cartilage	Fuchs endothelial corneal dystrophy
IX	$\alpha_1(IX)\alpha_2(IX)\alpha_3(IX)$	FACITs	Cartilage, Cornea, Vitreous	Multiple epiphyseal dysplasia
X	$[\alpha_1(X)]_3$	Network	Cartilage	Chondrodysplasia
XI	$\alpha_1(XI)\alpha_2(XI)\alpha_3(XI)$	Fibrillar	Cartilage, Intervertebral disc	Chondrodysplasia, osteoarthrosis
XII	$[\alpha_1(XII)]_3$	FACITs	Skin, Tendon	Bethlem myopathy
XIII	$[\alpha_1(XIII)]_3$	MACIT	Endothelial cells, Skin, Eye, Heart	–
XIV	$[\alpha_1(XIV)]_3$	FACITs	Bone, Skin, Cartilage, Eye, Nerve, Tendon	–
XV	$[\alpha_1(XV)]_3$	Multiplexin	Capillaries, Kidney, Heart, Testis	–
XVI	$[\alpha_1(XVI)]_3$	FACITs	Heart, Kidney, Skin	–
XVII	$[\alpha_1(XVII)]_3$	MACIT	Hemidesmosomes in epithelia	Epidermolysis bullosa
XVIII	$[\alpha_1(XVIII)]_3$	Multiplexin	Basement membrane, Liver	Knobloch syndrome

(continued)

Table 1.1 (continued)

Type	Composition	Subfamily	Distribution	Pathology
XIX	$[\alpha_1(XIX)]_3$	FACITs	Basement membrane	–
XX	$[\alpha_1(XX)]_3$	FACITs	Corneal epithelium, Embryonic skin, Sternal cartilage, Tendon	–
XXI	$[\alpha_1(XXI)]_3$	FACITs	Stomach, Kidney, Blood vessel walls, Secreted by smooth-muscle cells	–
XXII	$[\alpha_1(XXII)]_3$	FACITs	Tissue junctions	–
XXIII	$[\alpha_1(XXIII)]_3$	MACIT	Heart, Retina	Tumor
XXIV	$[\alpha_1(XXIV)]_3$	Fibrillar	Bone, Cornea	–
XXV	$[\alpha_1(XXV)]_3$	MACIT	Brain, Heart, Testis	Alzheimer's disease
XXVI	$[\alpha_1(XXVI)]_3$	Non-fibrillar	Testis, Ovary	–
XXVII	$[\alpha_1(XXVII)]_3$	Fibrillar	Cartilage	–
XXVIII	$[\alpha_1(XXVIII)]_3$	Beaded filament	Skin, Sciatic nerve	Neurodegenerative disease
XXIX	$[\alpha_1(XXIX)]_3$	Non-fibrillar	Suprabasal cells in epidermis, Lung, Small intestine, Colon and testis	–

1.3 Fibrillar Collagens

Fibrillar collagens, named for their ability to form fibrils, are the most abundant proteins in connective tissues [16]. They were the first members of the collagen superfamily to be discovered. Seven types of collagens, including type I, II, III, V, XI, XXIV, and XXVII, are capable of forming fibrils, and are thus classified as fibrillar collagens [17, 18]. By 1979, researchers had already identified five fibrillar collagens, including the most prevalent types I, II, and III, as well as two minor types V and XI [19]. Minor fibrillar collagen V is associated with major fibrillar collagen I, particularly in the cornea, while minor fibrillar collagen XI is observed in association with major fibrillar collagen II [20]. Only until 2003, two novel fibrillar collagens type XXIV and XXVII were discovered and found to be associated with tissues containing types I and II collagens, respectively [10, 21]. Hence, the five types of fibrillar collagens identified earlier are often referred to as "classical" fibrillar collagens. These five classical fibrillar collagens are composed of a common triple helix domain of similar lengths (1012–1029 amino acids). In contrast, the triple helix domain of novel fibrillar collagens type XXIV and XXVII is shorter (991–994 amino acids) and contains two interruptions in the repetitive $(Gly\text{-}X\text{-}Y)_n$ amino acid sequences.

Fibrillar collagen molecules can exist as homotrimers composed of three identical α chains, or heterotrimers consisting of two or three genetically distinct α chains [4, 20]. Individual α chains are denoted by the following nomenclature: $\alpha_n(N)$, where N represents the collagen type and n indicates the number of the α chain. Therefore, the chain composition of homotrimer collagen II is $[\alpha_1(II)]_3$, while that of heterotrimer collagen I is $[\alpha_1(I)]_2\alpha_2(I)$, indicating two identical α_1 chains and a third distinct α_2 chain [4]. For collagen V, it has three distinct α chains, and displays different chain compositions such as $[\alpha_1(V)]_2\alpha_2(V)$ and $\alpha_1(V)\alpha_2(V)\alpha_3(V)$ [22]. Collagen XI has three different α chains, while the $\alpha_3(XI)$ chain is nearly the same as $\alpha_1(II)$ [23]. Hybrid forms of collagen V and XI molecules have also been observed, further increasing the diversity of quantitatively minor fibrillar collagens [23, 24].

For all fibrillar collagens, the central large triple helical collagenous domain (COL1) is flanked by the N- and C-terminal domains called the N- and C-propeptides, respectively (Fig. 1.2) [7]. The C-propeptide forms a non-collagenous structure called the NC1 domain. The N-propeptide is more complex and can be divided into three sub-domains: a short non-collagenous linking sequence (NC2), a minor triple helical collagenous domain (COL2), and a globular non-collagenous N-terminal end (NC3) [18]. The N-propeptide exhibits the most structural variations among the three domains of classical fibrillar collagens, particularly in the NC3 sub-domain (Fig. 1.2). For example, the NC3 sub-domain of fibrillar collagen $\alpha_2(I)$ is relatively short, while the NC3 sub-domains of $\alpha_1(I)$, $\alpha_1(II)$, $\alpha_1(III)$, and $\alpha_2(V)$ chains contain a larger cysteine-rich repeat (CRR) domain. In contrast, the NC3 sub-domains of $\alpha_1(V)$, $\alpha_3(V)$, $\alpha_1(XI)$, $\alpha_2(XI)$, $\alpha_1(XXIV)$, and $\alpha_1(XXVII)$ can be divided into two parts: a thrombospondin N-terminal-like (TSPN) domain and a variable region (VR) [25].

Fibrillar collagens are initially synthesized as procollagens containing both N- and C-propeptides in connective tissue cells, and then secreted into the extracellular space (Fig. 1.3) [5, 17]. The N- and C- propetides of procollagens are thereafter cleaved by proteolytic enzymes to achieve a mature and functional form of collagen. The removal of propeptides is considered as a critical factor to trigger the fibril formation for the classical fibrillar collagens, whose fibrils are subsequently stabilized by the formation of covalent cross-links. However, an exception is observed for minor fibrillar collagens V and XI that their mature molecule maintains a part of the N-terminal domain, which has been proposed to mediate the heterotypic growth by sterically affecting lateral interaction.

1.4 Nonfibrillar Collagens

The fibrillar collagens are primarily composed of a major triple-helical domain, while nonfibrillar collagens can consist of several triple-helical domains interspersed with non-collagenous domains that contribute to structural assembly and biological activities [7, 17]. In 1991, the discovery of collagen IX demonstrated for the first time that collagenous domains could be interrupted by non-collagenous domains.

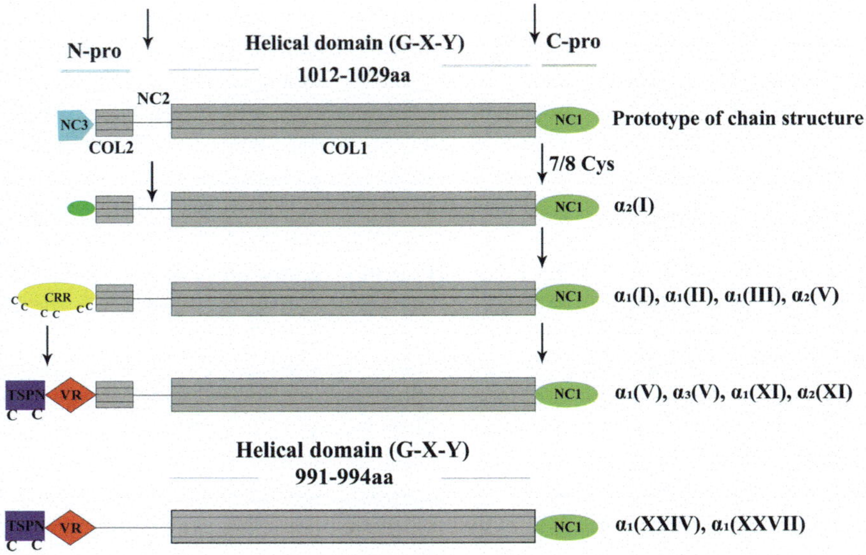

Fig. 1.2 Domain structures of fibrillar collagens

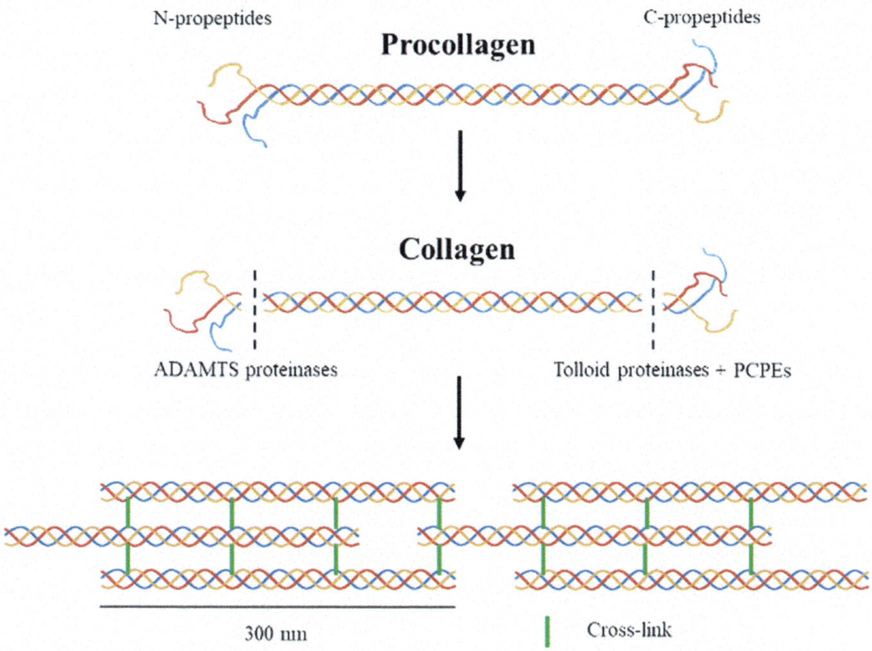

Fig. 1.3 Fibril assembly of fibrillar collagens

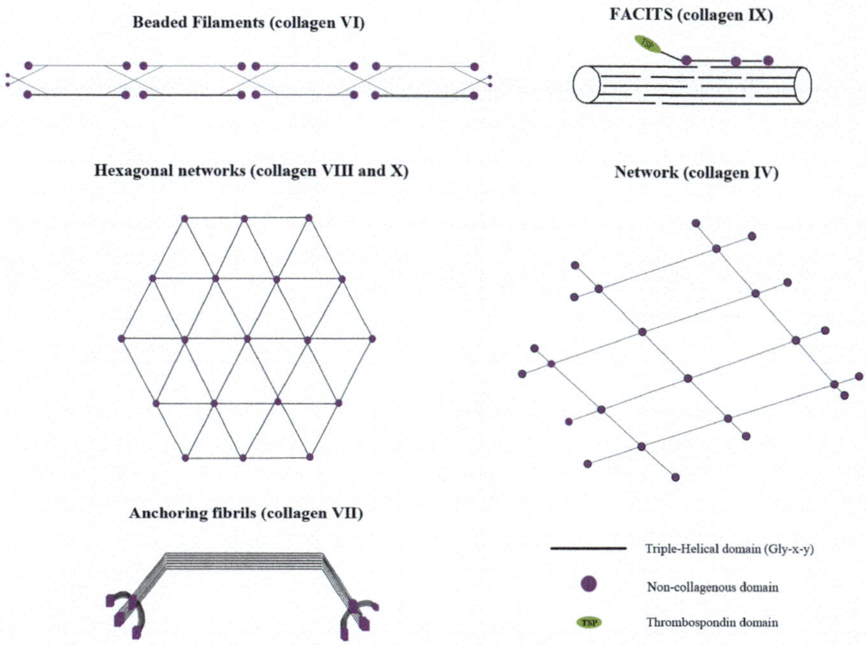

Fig. 1.4 Supramolecular assemblies of nonfibrillar collagens

Nonfibrillar collagens are categorized into subfamilies based on their supramolecular assemblies, including FACITs (Fibril Associated Collagens with Interrupted Triple Helices), Network, Beaded filament, Anchoring collagen, Multiplexins (Multiple Triple Helix Domains with Interruptions), MACIT (Membrane Associated Collagens with Interrupted Triple Helices), and other types (XXVI and XXIX) (Fig. 1.4).

1.4.1 FACITs

The FACITs represent the largest group within the collagen suprafamily, comprising eight distinct types. While they are incapable of forming fibrils on their own, they are instead found associated with the surface of collagen fibrils [7]. Collagen IX is a founding member of FACITs, and plays a crucial role in cartilage collagen fibrils. It is an $\alpha_1(IX)\alpha_2(IX)\alpha_3(IX)$ heterotrimer, which consists of three short collagenous domains and four non-collagenous domains [26]. Collagen IX covalently cross-links to collagens II and XI on the surface of cartilage fibrils, where it serves as a critical component [27]. Collagen XII is a homotrimer composed of three identical α_1(XII) chains, each comprising two collagenous domains and three non-collagenous domains [28]. The molecule is found on the surface of collagen fibrils and is localized to collagen-I rich connective tissues.

1.4.2 Network

Type IV collagen is the most abundant network collagen found in basement membranes [29, 30]. It forms a meshwork-like structure that acts as a filtration barrier in the kidney glomerulus and provides stability and mechanical support to blood vessels [7]. Each chain of type IV collagen contains over 20 interruptions in the repetitive (Gly-X-Y)$_n$ sequence, and there are six known chains that form three heterotrimeric molecules: $[\alpha_1(IV)]_2\alpha_2(IV)$, $\alpha_3(IV)\alpha_4(IV)\alpha_5(IV)$ and $[\alpha_5(IV)]_2\alpha_6(IV)$ [29]. Collagen IV molecules form tetramers through the assembly of four molecules via their N-terminal 7S domain, and also form NC1 dimers when two molecules associate through their C-terminal NC1 domain. These processes, along with triple-helical interactions and interactions with other basement membrane components, facilitate the formation of an open meshwork structure of collagen IV [31]. The stability of the assembled supramolecular structure of collagen IV is further enhanced by disulfide bonding in the 7S region and lysine-derived cross-links. Collagens VIII and X, on the other hand, associate to form hexagonal lattice-like networks in Descemet's membrane and hypertrophic cartilage, respectively [4].

1.4.3 Anchoring Collagen

Collagen VII is a key component of anchoring fibrils located in the basement zone beneath stratified squamous epithelia [32, 33]. It consists of three identical α chains and contains the longest triple-helical region with interruptions of approximately 420 nm in length. The collagen molecule is flanked by non-collagenous NC1 and NC2 domains at the N- and C-termini, respectively [32]. Collagen VII is produced by both epidermal keratinocytes and dermal fibroblasts, and forms anti-parallel dimers with N-terminal domains present at both ends of the molecule. The dimer assembly is initiated by proteolytic processing of the NC2 domain and further stabilized by the formation of intermolecular disulfide bonds. Collagen VII dimers assemble laterally in large numbers to form anchoring fibrils, which provide stability to the dermal-epidermal adhesion at the dermal site of the lamina lucida/papillary dermis interface [33]. The enhanced stability is likely attributed by the high affinity of the NC1(VII) domain to bind collagen IV. Mutations in collagen VII can result in severe blistering disorders called epidermolysis bullosa [34].

1.4.4 Beaded Filament

Collagen VI is a unique collagen, consisting of short triple-helical regions (about one-third the length of fibrillar collagens) and large N- and C-terminal regions containing mainly von Willebrand factor A domains [35]. It is a heterotrimer

$\alpha_1(VI)\alpha_2(VI)\alpha_3(VI)$, and assembles into beaded filaments with a periodicity of 110 nm [4, 36]. Collagen VI, unlike fibrillar collagens, does not undergo any enzymatic processing of its N- and C-terminal non-collagenous regions. The supramolecular assembly of collagen VI begins inside the cell with the formation of antiparallel dimers and tetramers, which further assemble in the extracellular matrix to form beaded filaments or microfibrils. Collagen VI plays a critical role in maintaining tissue integrity, and its defects have been associated with various genetic disorders, including Bethlem myopathy and Ullrich congenital muscular dystrophy [37].

1.4.5 Multiplexins

Collagens XV and XVIII are classified as Multiplexins due to the presence of multiple triple-helical domains, with 9 domains in collagen XV and 10 domains in collagen XVIII [17, 38]. Collagens XV and XVIII are homotrimers with common C-terminal functional domains, known as endostatins, which share approximately two-thirds sequence identity. They both contain a thrombospondin-1 domain at the N-terminal region, which is homologous to the heparin-binding domain of thrombospondin-1. In addition, they are both proteoglycans, with collagen XV containing chondroitin sulfate chains and collagen XVIII containing heparan sulfate chains. The $\alpha_1(XV)$ collagen chain contains a large N-terminal non-triple helical domain with a tandem repeat structure consisting of a 45-amino acid residue sequence. Collagen XV is believed to play a critical role in stabilizing skeletal muscle cells and micro-vessels [39]. In contrast, collagen XVIII is predominantly localized in the perivascular basement membrane, where it is considered a key component for maintaining the structural integrity of basement membranes [40]. These two collagens are primarily found in different tissues (collagen XV in the muscle and collagen XVIII in the eye), and are believed to have distinct biological functions [39].

1.4.6 MACIT

Collagens XIII, XVII, XXIII, and XXV belong to the family of membrane-associated collagens with interrupted triple helices (MACIT), and they all share a single-pass hydrophobic transmembrane domain [17, 41]. In contrast to fibrillar collagens, collagens XIII and XVII fold from N- to C-terminus. Collagen XIII is a homotrimer composed of a short intracellular domain, a single transmembrane region, and an extracellular ectodomain with three collagenous domains interspersed by short non-collagenous domains [42]. Collagen XIII is considered to play a critical role in epidermal cell–matrix and cell–cell interactions in cultured keratinocytes. Collagen XVII is a transmembrane collagen and a major component of the hemidesmosome, with its extracellular domain composed of 15 collagenous domains separated by 16 non-collagenous domains [43]. Collagen XXIII is also a transmembrane collagen but is primarily found in metastatic tumor cells, and it consists of an N-terminal cytoplasmic domain, a transmembrane domain, and three collagenous domains interspersed with short non-collagenous domains [44]. Collagen XXV, on the other hand, is a short collagen composed of three collagenous domains flanked by four non-collagenous domains [45]. MACIT collagens function as cell surface receptors, and their ectodomains are cleaved at the extracellular face of the membrane. These ectodomains retain their triple helical structure after release into the extracellular space and participate in interactions with other components of the extracellular matrix.

1.4.7 Type XXVI Collagen

Type XXVI collagen is specifically expressed in testicular and ovarian tissues and consists of two collagenous domains of 69 and 33 amino acids and three non-collagenous domains [46]. The trimerization of type XXVI collagen polypeptide chains is initiated by the formation of intermolecular disulfide bonds within the N-terminal non-collagenous domain. Type XXVI collagen differs from FACITs in that it lacks the conserved motif containing two cysteine residues at the COL1-NC1 junction, as well as a thrombospondin N-terminal-like domain. Instead, it contains two coiled-coil oligomerization domains similar to those found in membrane collagens, but lacks a transmembrane sequence. It is worth noting that the ectodomain of type XXVI collagen cannot be cleaved by furin convertase despite the presence of a consensus furin protease cleavage site [46].

1.4.8 Type XXIX Collagen

Type XXIX collagen is the newest member of the collagen family and is primarily found in the tissues of the epidermis, lung, small intestine, colon, and testis [4]. It is a large protein composed of 2,614 amino acids with a calculated molecular mass of 289.9 kDa, and shares 32% amino acid identity with COL6A3. Collagen XXIX has a short and uninterrupted triple helical region flanked by 6 N-terminal and 3 C-terminal von Willebrand factor A domains. These domains are involved in protein–ligand interactions for the organization of tissue architecture and cell adhesion.

1.5 Biological Interactions of Collagen

Collagen is not only the major structural protein in the extracellular matrix, but also plays a crucial role in many cellular processes, such as cell development, adhesion, and migration [47]. Cells have evolved diverse mechanisms to interact with collagen, including adhesion to collagen fibers, response to injury by contacting collagen, and migration through the collagen-rich stroma during normal morphogenic processes such as mammary gland branching and pathological conditions such as tumor metastasis. These situations have led cells to employ a broad range of proteins to interact with collagen, despite its relatively uniform rod-like conformation and repeating $(Gly\text{-}X\text{-}Y)_n$ sequences. Collagen can bind specifically to a plethora of molecules such as integrins, discoidin domain receptor, glycoprotein VI, leukocyte-associated IG-like receptor-1, the mannose receptor family, and matrix metalloproteinases (Fig. 1.5) [48, 49].

1.5.1 Integrins

Integrins are the primary collagen receptors that mediate cell adhesion to the extracellular matrix. By regulating cell–cell and cell–matrix contacts, integrins play essential roles in a wide range of biological interactions, including cell development and differentiation, tissue morphogenesis, wound healing, angiogenesis, inflammation, innate and antigen-specific immunity, and hemostasis [50, 51]. Dysregulation of integrins may lead to various pathological conditions ranging from autoimmunity, thrombotic vascular diseases to cancer metastasis. Integrins are heterodimers consisting of α and β subunits, with each comprising a large extracellular domain, a single transmembrane domain, and a short cytoplasmic domain. Structural rearrangements of these domains enable integrins to transmit signals bidirectionally across the plasma

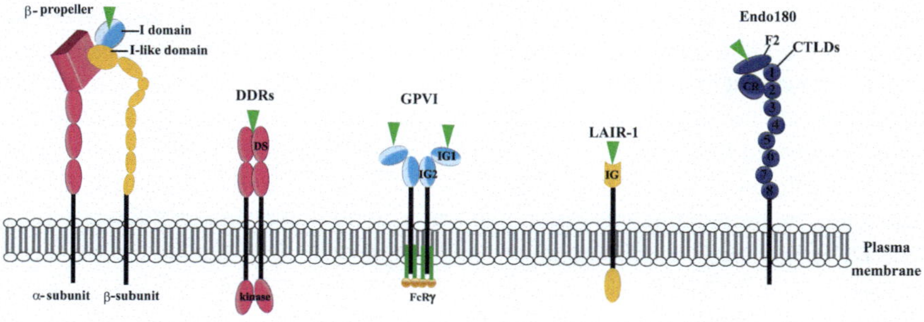

Fig. 1.5 Schematic domain structures of mammalian collagen receptors. The collagen-binding sites are indicated by green triangles

membrane [52]. There are 18 α and 8 β subunits that combine to form at least 24 distinct integrins.

The extracellular domain of the integrin α subunit consists of a seven-bladed β-propeller with an inserted I domain, followed by three immunoglobulin-like (IG) domains. Meanwhile, the extracellular domain of the β subunit comprises a domain related to the α subunit I domain, an IG-like hybrid domain, a plexin–semaphorin–integrin (PSI) domain, four epidermal growth factor-like (EGF) domains, and a β-tail domain [53]. The I-domain in the α subunit of integrin (also called the A domain due to its homology to von Willebrand factor A domains) is located between the blade 2 and 3 of the β-propeller, and has been identified as the major ligand-binding domain. The isolated I-domains retain the specificity and high affinity of their parent integrins towards collagen. Integrins can be categorized into subgroups based on their ligand-binding properties [54]. Among the integrins, four β_1 integrins with an I-domain ($\alpha_1\beta_1$, $\alpha_2\beta_1$, $\alpha_{10}\beta_1$, and $\alpha_{11}\beta_1$) primarily function as collagen receptors, while $\alpha_1\beta_1$ and $\alpha_2\beta_1$ are the most widely distributed [55]. The I domains of integrins α_1 and α_2 share almost the same secondary structure, while they possess different selectivity to collagen subtypes.

Integrins $\alpha_2\beta_1$ and $\alpha_1\beta_1$ recognize and bind to fibrillar collagens through a critical GXX'GER hexapeptide motif [56]. A number of GXX'GER sequences in collagen have been identified, showing very different binding affinities towards integrins. Among these sequences, the GFOGER sequence exhibits the highest affinity and can activate resting platelets by "outside-in" signaling. In contrast, low to intermediate-affinity sequences are activated by "inside-out" signaling, indicating their biological relevance [54]. The comparison of X-ray crystal structures of free triple helical peptide $(GPO)_2GFOGER(GPO)_3$ and the peptide complexed with the I-domain of integrin $\alpha_2\beta_1$ reveals that the key GFOGER sequence in collagen undergoes substantial conformational changes during the binding process to integrin.

1.5.2 DDRs

The discoidin domain receptors (DDRs) are a unique subfamily of receptor tyrosine kinases (RTKs) that serve as collagen receptors. Unlike most RTKs, which are activated by growth factors, DDRs can be activated by extracellular matrix components. DDR1 is predominantly present in epithelial cells and leukocytes, while DDR2 is mainly expressed in mesenchymal cells [54]. Both DDR1 and DDR2 play crucial roles in regulating cell adhesion, migration, proliferation, and extracellular matrix remodeling by modulating the expression and activity of matrix metalloproteinases [48]. DDR homologs are homodimers and single-span transmembrane proteins, consisting of an extracellular domain with an N-terminal discoidin (DS) domain, followed by a DS-unique sequence with approximately 200 amino acids, and a cytoplasmic domain comprising a juxtamembrane domain and the C-terminal catalytic tyrosine kinase domain.

The discoidin domain of DDRs has been identified as the collagen-binding domain. In vitro affinity experiments using recombinant DDR extracellular domains have demonstrated that the discoidin domain alone is sufficient to strongly bind collagen [57]. The ability of DDRs to bind collagen requires dimerization and proper arrangement of the discoidin domains within the dimer. Monomeric DDRs or misaligned dimeric DDRs cannot bind collagen. Additionally, the collagen-binding sites in DDRs are suggested to be three spatially adjacent surface-exposed loops in the discoidin domain. Both DDR1 and DDR2 are known to recognize a variety of fibrillar and non-fibrillar collagens, but they exhibit distinct preferences for certain types [58]. DDR1 is considered the preferred receptor for collagen IV, while DDR2 is known to have a stronger affinity for collagen II and collagen X. Notably, both DDR1 and DDR2 only bind to native collagen in its triple helical state and not in its denatured state. Further studies on the binding between the discoidin domain and collagen have suggested that the binding sequences on collagen span only two or three triplets and contain some charged residues [59].

1.5.3 GPVI

Glycoprotein VI (GPVI) is a platelet membrane glycoprotein receptor for collagen that plays a crucial role in the collagen-induced activation and aggregation of platelets [60]. GPVI-deficient platelets lack specific collagen-induced aggregation and lose the ability to form thrombi on collagen surfaces under flow conditions, demonstrating the essential role of GPVI in collagen-induced platelet activation [61]. The GPVI-collagen interaction is considered a key step in triggering intracellular signaling, resulting in the release of aggregation mediators such as ADP and thromboxane A2 [62]. Platelet adhesion requires the direct binding of collagen not only to GPVI but also to the platelet receptor $\alpha_2\beta_1$ integrin. GPVI and $\alpha_2\beta_1$ integrin are proposed to function in a highly cooperative and integrated fashion to mediate firm adhesion of platelets, thrombus formation and procoagulant activity.

GPVI contains two immunoglobulin-like (IG) domains that are heavily O-glycosylated mucin-like stalk regions, a transmembrane helix and a short cytoplasmic domain that binds calmodulin and SH3 domains [63, 64]. The first IG domain (IG1) has been identified as the collagen-binding site, as revealed by site-directed mutagenesis. It is suggested that two GPVI molecules form a complex with one Fc receptor (FcR) γ-chain dimer to act as a functional receptor on the platelet surface. The formation of such a dimeric complex is a prerequisite for GPVI to exhibit high affinity binding to collagen. Furthermore, the tyrosine phosphorylation of the immunoreceptor tyrosine-based activation motif (ITAM) of the FcR γ-chain is a key factor in triggering GPVI-induced activation [65].

1.5.4 LAIR-1

Leukocyte-associated IG-like receptor-1 (LAIR-1) is a collagen receptor with the simplest structure, consisting of a single IG domain, a transmembrane helix, and a short cytoplasmic domain containing two immunoreceptor tyrosine-based inhibitory motifs (ITIMs) [66]. LAIR-1 was first found to bind collagen XVII, and later found to bind collagens I and III as well. LAIR-1 also specifically recognizes $(GPO)_{10}$, but not $(GPP)_{10}$, showing a similarity to GPVI in terms of binding preferences [67]. LAIR-1 is an inhibitory receptor found on peripheral mononuclear cells, which regulates the immune response to prevent lysis of cells recognized as self [68].

1.5.5 The Mannose Receptor Family

The mannose receptor family consists of four glycoprotein members: mannose receptor (MR), the M-type phospholipase A2 receptor (PLA2R), DEC-205, and Endo180 [69]. They are all type I transmembrane receptors sharing a common domain architecture with an N-terminal cysteine-rich (CR) domain, a single fibronectin type II (F2) domain and eight to ten C-type lectin-like domains (CTLDs) [70]. All the four family members have been observed to control the uptake of various extracellular ligands, such as glycoproteins (MR), soluble phospholipases (PLA2R), antigens (DEC-205), and collagen (PLA2R, Endo180, MR). Endo180 is known for its broad collagen specificity and binds avidly to denatured collagen (gelatin) [71].

1.5.6 Matrix Metalloproteinase

The matrix metalloproteinase (MMP) family comprises at least 25 members and is responsible for the catabolism of extracellular matrix (ECM) components. Collagen, with its triple helix structure, is resistant to most proteases except for a few proteinases including MMP-1, MMP-2, MMP-8, MMP-13, MMP-14, MMP-18, and MMP-22 [72]. The binding and cleaving of collagen by these MMPs are critical steps in collagen degradation, which is essential in biological processes such as wound healing, cell migration, tissue remodeling, and organ morphogenesis. However, accelerated breakdown of collagen may result in diseases such as arthritis, tumor cell invasion, glomerulonephritis, and metastasis [73]. These MMPs cleave interstitial collagens at a unique locus approximately three-fourths away from the N-terminus of the collagens, after a Gly residue in the sequence of Gly-[Ile/Leu]-[Ala/leu]. Interestingly, dozens of other sites in the collagens contain the same cleavage scissile bond Gly-Ile/Leu but are not hydrolyzed. It has been suggested that the distribution of 'weak helix' triplets, imino acids, hydroxyproline, and charged residues as well as the side-chain volume concentration may provide recognition information for MMPs.

1.5.7 *Other Collagen-Binding Biomolecules*

Cluster of differentiation 44 (CD44) is a cell-surface glycoprotein that functions as a receptor with a high affinity for collagen. It is often overexpressed in metastatic tumors and is considered a marker for cancer stem cells [74]. CD44-expressing cells within tumor populations maintain the vitality of the tumor by self-renewal and unlimited proliferation. In metastatic melanoma cells, CD44 is particularly upregulated. Type IV collagen can be used to target the CD44 receptor (α_1-1263–1277: GVKGDKGNPGWPGAP) [75]. The interaction between CD44 and collagen requires the triple helix conformation of the collagen molecule.

Collagen has the ability to bind to von Willebrand factor (vWF) and heparan sulfate proteoglycans (HSPGs) in addition to its interaction with transmembrane receptors. The binding of collagen to vWF is crucial in regulating the speed of platelets through blood circulation. The collagen-binding domain of vWF, with the sequence GPRGQOGVMGFO, is found in both type II and type III collagen [76]. HSPGs, which contain a high density of negative charges, serve as excellent ligands for binding growth factors [77]. Synthetic collagen scaffolds that contain heparin-binding epitopes can bind soluble growth factors, such as vascular endothelial growth factor, due to the presence of heparin. Thus, incorporating heparin-binding epitopes into collagen scaffolds increases their biological functions.

1.5.8 *Collagen Binding Proteins with High-Resolution Structures*

A diverse group of cell surface receptors have been identified to specifically interact with triple-helical collagen, despite the limiting scope for specific interactions due to the repeating nature and high imino acid content of collagen triple helices. Discriminating between fibrillar collagen, isolated triple helices, and denatured collagen (gelatin) presents demanding challenges for collagen receptors [49]. However, these receptors have overcome these constraints and achieved remarkable specificity. Integrins and DDRs likely recognize specific amino acid side chains in collagen, while glycoprotein VI and LAIR-1 probably recognize the bare triple helix of uninterrupted GPO repeats. The mannose receptor family recognizes both native collagen and gelatin. Deciphering the complex modes of collagen interactions with different receptors will contribute to a better understanding of the biological roles of collagen. Collagen binding proteins with solved high-resolution structures are summarized in Table 1.2.

Table 1.2 Collagen binding proteins with high-resolution structures. PDB entries are shown for the structures of the proteins alone and the complex of the proteins with triple-helical peptide (THP)

Collagen binding molecules	PDB ID		Binding sequence
	Alone	+ THP	
Integrin $\alpha_1\beta_1$	1PT6	2M32	GFOGEN, GFOGER
Integrin $\alpha_2\beta_1$	1AOX	4BJ3	GFOGER
DDR1	4AG4	–	GVMGFO
DDR2	2Z4F	2WUH	GVMGFO
GPVI	2GI7	–	GAOGLRGGAGPOGPEGGKGAAGPOGPO
LAIR-1	3KGR	–	Multiple sites
Fibronectin	3M7P	3EJH	GLAGQRGIVGLPGQRGER
PEDF	1IMV	–	KGHRGFSGL
SPARC	1BMO	2V53	GPSGPRGQOGVMGFOGPK
VWF	1AO3	4DMU	RGQOGVMGF
MMP-1	2CLT	4AUO	GPQGLAGQRGIVGLP
Hsp47	4AU4	4AU3	xGxR
P4H	1TJC	4BTA	–
Decorin	1XKU	–	Multiple sites
Heparin	3IRL	–	KGHRGF

1.6 Collagen Diseases

Collagen is defined by a repeating pattern of $(\text{Gly-X-Y})_n$ in its triple helix domains, which is maintained in fibrillar collagens but may contain natural interruptions in nonfibrillar collagens [78]. Thousands of mutations have been identified in 12 out of 28 types of collagens, and even a single Gly mutation in the triple helix domain of either fibrillar or nonfibrillar collagens can lead to tissue-level diseases. For instance, more than 600 different Gly to X mutations in type I collagen have been found to be associated with osteogenesis imperfecta (OI), a connective tissue disorder that results in fragile bones and a high risk of fractures [79]. Gly to X mutations in type II collagen are linked to Type II collagenopathies, characterized by abnormal development of bone; while Gly substitutions in type III collagen lead to Ehlers Danlos Syndrome type IV, which may result in sudden death due to aortic ruptures. Other collagen-related diseases include Alport syndrome, osteoporosis, Knobloch syndrome, and more [80].

1.6.1 *Osteogenesis Imperfecta*

Osteogenesis Imperfecta (OI), commonly known as "brittle bone disease", is a heritable connective tissue disorder with a prevalence of approximately 1:15,000 births [81]. The main cause of OI is the defect in type I collagen, including abnormal primary structure, insufficient quantity, irregular post-translational modification, and misfolding [79]. This results in the major clinical features of OI, which are low bone mass and reduced bone material strength, leading to various phenotypes such as bone fragility, bone deformity, and growth deficiency [82–84]. OI is a heterogeneous disorder with broad phenotypes and can be classified into four groups based on clinical, radiologic, and genetic criteria (Table 1.3). The mildest form is OI type I, characterized by mild to moderate bone fragility, blue sclerae, and premature deafness [85–87]. The most severe form is OI type II, resulting in perinatal lethality with affected infants experiencing intrauterine fractures and intracranial hemorrhage after vaginal delivery, passing away shortly after birth [87]. OI Type III is the most severe, non-lethal form that is characterized by progressive deformities, while affected individuals with type III OI have multiple fractures, extremely short stature, triangular facies, frontal bossing, vertebral compressions and scoliosis [87]. OI Type IV is a moderately severe form of OI, and has a broad phenotypic range, overlapping mild types I and III. Affected individuals with type IV OI may exhibit moderate bowing of long bones, variable short stature, osteoporosis and mild scoliosis [87].

Osteogenesis Imperfecta (OI) is predominantly inherited in an autosomal dominant manner, and more than 1500 mutations have been identified in the genes encoding the α_1(I) and α_2(I) chains of type I collagen (COL1A1 and COL1A2), resulting in various phenotypes of OI by altering the structure or quantity of type I collagen [86]. More recently, autosomal recessive forms of OI have been identified, and are caused by defects in genes whose products interact with type I collagen, rather than primary defects in type I collagen itself. Most recessive OI cases are caused by deficiency of any of the three components of the collagen prolyl 3-hydroxylation complex (CRTAP, LEPRE1 and PPIB), which can lead to delayed collagen folding and increased modification of the collagen helix due to the absence of 3-hydroxylation [88]. Other causes of recessive OI include absence or dysfunction of two collagen

Table 1.3 Classification of OI phenotypes. AD: Autosomal dominant; AR: Autosomal recessive [79, 86, 87]

OI type	Inheritance	Clinical characteristics
I	AD	Normal stature, Little or no deformity, Blue sclerae, Hearing loss
II	AD	Lethal in perinatal period, Beaded ribs, Long bone fractures
III	AD/AR	Progressively deforming, Short stature, Multilpe fractures, Triangular face, Hearing loss
IV	AD	Moderately severe, Variable short stature, Dentinogenesis imperfecta, Osteoporosis, Bowing of long bones

chaperones, FKBP65 and HSP47, which are involved in the triple helix folding and secretion of collagen [89].

1.6.2 Collagenopathies

The type II and XI collagenopathies are a group of inheritable connective tissue disorders caused by defects in type II or type XI collagen. Type II and type XI collagen disorders are grouped together since both types of collagen are components of the cartilage found in joints, spinal column, and inner ear [2]. The clinical presentation of collagenopathies is highly variable, ranging from perinatal lethality to mild conditions, including achondrogenesis type II, hypochondrogenesis, Kniest dysplasia, spondyloepiphyseal dysplasia congenita, spondyloperipheral dysplasia, and Stickler syndrome. Common symptoms include short stature, enlarged joints, and spinal curvature [80].

Mutations in the COL2A1, COL11A1 and COL11A2 genes can result in structural defects or deficient quantity of collagen II and XI, which can interfere with the normal development of bones and other connective tissues, leading to type II and XI collagenopathies. The nature and position of the mutations within the polypeptide chain are found to be correlated with the severity of the collagenopathies. Mutations that interfere with the folding of the triple helix and are located near the C-terminal end are associated with the most severe phenotypes, such as achondrogenesis type II. On the other hand, premature stop codons leading to reduced quantity of normal collagen II tend to produce moderate phenotypes, such as Stickler syndrome [90].

1.6.3 Ehlers-Danlos Syndrome

Ehlers-Danlos syndrome (EDS) is a heterogeneous group of connective tissue diseases that affects roughly 1 in 5,000 individuals and is characterized by clinical features such as skin hyperextensibility, joint hypermobility, and tissue fragility [91]. The main cause of EDS is defect in the structure, production, or processing of collagen or proteins that interact with collagen. The phenotypes of EDS vary widely, from mildly loose joints to life-threatening complications such as aortic dissection. EDS has been classified into six primary subtypes using a simplified, revised protocol based upon the underlying biochemical cause, mode of inheritance, major symptoms, and physical findings.

The EDS classical type (formerly EDS I and II) is associated with mutations in the COL5A1 and COL5A2 genes and is characterized by excessive laxity and extension of the joints, susceptibility to recurrent sprains and dislocations of certain joints, abnormally increased elasticity and extension of the skin, and abnormal healing of skin wounds. EDS hypermobility type (formerly EDS III) is primarily characterized by generalized, excessive extension of the large and small joints and is autosomal dominant, while a specific underlying collagen defect has not been identified [54]. EDS vascular type (formerly EDS IV) is the most severe type due to its frequent lethal complications such as arterial rupture and is primarily characterized by unusually thin, transparent skin and a susceptibility to severe bruising from minor trauma [92]. Mutations in the COL3A1 gene lead to abnormal production of collagen III, resulting in EDS vascular type [91].

EDS kyphoscoliosis type (formerly EDS VI) is caused by mutations in the PLOD gene encoding a collagen-modifying enzyme known as lysyl hydroxylase [80]. It shows primary symptoms including abnormal sideways curvature of the spine, diminished muscle tone, and excessive extension and looseness of the joints. EDS arthrochalasia type (formerly EDS VII, Autosomal Dominant) is caused by mutations in the COL1A1 and COL1A2 genes, and it is mainly characterized by dislocation of the hips at birth, severe, excessive extension of the joints, and recurrent partial dislocations of affected joints. EDS dermatosparaxis type (formerly EDS VII, Autosomal Recessive) is suggested to result from mutations of a gene encoding a collagen-modifying enzyme known as procollagen I N-terminal peptidase [93]. It displays symptoms such as severe skin fragility, sagging, redundant skin, and extensive bruising.

1.6.4 Alport Syndrome

Alport syndrome is an inherited disorder with a prevalence of 1 in 50,000 newborns, characterized by glomerulonephritis, end-stage kidney disease, hearing loss, and eye abnormalities [94]. It is caused by mutations in the genes COL4A3, COL4A4, and COL4A5, which encode basement membrane collagen type IV and interfere with the proper production or assembly of the type IV collagen in the glomerulus. The absence of the collagen IV network from basement membranes is first to be found to be associated with hematuria [95]. Since type IV collagen is widely present in the ears, eyes, and kidneys, Alport syndrome affects seemingly unrelated parts of the body such as ears, eyes, and kidneys. Approximately 85% of patients inherit Alport syndrome in an X-linked pattern resulting from mutations in the COL4A5 gene. Therefore, affected males frequently experience severe Alport syndrome and develop eventual kidney failure, while affected females often suffer from relatively mild Alport syndrome such as blood in the urine. Mutations in the COL4A3 and COL4A4 genes are associated with autosomal recessive inheritance, while heterozygous mutations may cause autosomal dominant disease [96].

1.6.5 Osteoporosis

Osteoporosis is a skeletal disorder characterized by low bone mass and microarchitectural deterioration of bone tissue, which predisposes the individual to an increased risk of fractures [80, 97]. Osteoporosis has become a major public health problem, and is still growing more widespread with the aging of the world population. It is widely accepted that genetic factors play critical roles in the pathogenesis of osteoporosis. Type I collagen, the key protein component of bone, is encoded by the COLIA1 and COLIA2 genes, while a G– > T polymorphism at a recognition site for the transcription factor Sp1 in COLIA1 has been found to be significantly correlated with bone mass and osteoporotic fracture [78, 98]. G/T heterozygotes at the Sp1 site show much lower bone mineral density (BMD) than G/G homozygotes, making the COLIA1 Sp1 polymorphism an important marker for low BMD and vertebral fracture. The COLIA1 Sp1 polymorphism has been shown to potentially predict the response to bisphosphonate therapy in the treatment of osteoporosis.

Bone mineral density (BMD) can be measured by dual X-ray absorptiometry (DXA), which is the most commonly used method to diagnose osteoporosis. However, BMD measurement only captures some risk factors for fracture. Noninvasive detection of serum and urinary biochemical markers provides critical information about quantitative changes in skeletal turnover, and it complements the BMD measurement to assess the risk of osteoporosis. Post-translational modifications in type I collagen have led to the development of key biochemical markers reflecting changes in the material property of bone, an important determinant of bone strength [78, 80, 98]. Serum procollagen type I N-terminal propeptide (PINP) and the crosslinked C-terminal telopeptide (serum CTX) have been widely utilized as reference markers of bone formation and bone resorption, respectively.

References

1. Miller EJ, Gay S (1987) The collagens: an overview and update. Methods Enzymol 144:3–41
2. Shoulders MD, Raines RT (2009) Collagen structure and stability. Annu Rev Biochem 78:929–958
3. Ricard-Blum S (2011) The collagen family. Cold Spring Harb Perspect Biol 3:a004978
4. Sorushanova A, Delgado LM, Wu ZN et al (2019) The collagen suprafamily: from biosynthesis to advanced biomaterial development. Adv Mater 31:e1801651
5. Peter F (2008) Collagen: structure and mechanic. Springer, Germany
6. Walker DR, Hulgan SAH, Peterson CM et al (2021) Predicting the stability of homotrimeric and heterotrimeric collagen helices. Nat Chem 13:260–269
7. Brinckmann J (2005) Collagens at a glance, vol 247. Springer, Heidelberg, pp 1–6
8. Kielty CM, Grant ME (2003) Connective tissue and its heritable disorders. In: The collagen family: structure, assembly, and organization in the extracellular matrix, 2nd edn. Wiley, New York, pp 159–221
9. Kielty CM, Grant ME (2002) The collagen family: structure, assembly, and organization in the extracellular matrix. In: Connective tissue and its heritable disorders, pp 159–221
10. Pace JM, Corrado M, Missero C et al (2003) Identification, characterization and expression analysis of a new fibrillar collagen gene, COL27A1. Matrix Biol 22:3–14
11. Wang C, Kovanen V, Raudasoja P et al (2009) The glycosyltransferase activities of lysyl hydroxylase 3 (LH3) in the extracellular space are important for cell growth and viability. J Cell Mol Med 13:508–521
12. Soderhall C, Marenholz I, Kerscher T et al (2007) Variants in a novel epidermal collagen gene (COL29A1) are associated with atopic dermatitis. PLoS Biol 5(9):e242

13. Culav EM, Clark CH, Merrilees MJ (1999) Connective tissues: matrix composition and its relevance to physical therapy. Phys Ther 79:308–319
14. Wu Z, Korntner SH, Mullen AM et al (2021) Collagen type II: from biosynthesis to advanced biomaterials for cartilage engineering. Biomater Biosyst 4:100030
15. Aziz J, Shezali H, Radzi Z et al (2016) Molecular mechanisms of stress-responsive changes in collagen and elastin networks in skin. Skin Pharmacol Physiol 29:190–203
16. Sun B (2021) The mechanics of fibrillar collagen extracellular matrix. Cell Rep Phys Sci 2:100515
17. Ramachandran GN, Reddi AH (1976) Biochemistry of collagen. Plenum Press, New York, pp 4–545
18. Ricard-Blum S, Ruggiero F (2005) The collagen superfamily: from the extracellular matrix to the cell membrane. Pathol Biol (Paris) 53:430–442
19. Burgeson RE, Hollister DW (1979) Collagen heterogeneity in human cartilage: identification of several new collagen chains. Biochem Biophys Res Commun 87:1124–1131
20. Richards AJ, Snead MP (2022) Molecular basis of pathogenic variants in the fibrillar collagens. Genes (Basel) 13:1199
21. Koch M, Laub F, Zhou P et al (2003) Collagen XXIV, a vertebrate fibrillar collagen with structural features of invertebrate collagens: selective expression in developing cornea and bone. J Biol Chem 278:43236–43244
22. Mizuno K, Bachinger HP, Imamura Y et al (2013) Fragility of reconstituted type V collagen fibrils with the chain composition of alpha1(V)alpha2(V)alpha3(V) respective of the D-periodic banding pattern. Connect Tissue Res 54:41–48
23. Wu JJ, Weis MA, Kim LS et al (2009) Differences in chain usage and cross-linking specificities of cartilage type V/XI collagen isoforms with age and tissue. J Biol Chem 284:5539–5545
24. Kypreos KE, Sonenshein GE (1998) Basic fibroblast growth factor decreases type V/XI collagen expression in cultured bovine aortic smooth muscle cells. J Cell Biochem 68:247–258
25. Fang M, Jacob R, McDougal O et al (2012) Minor fibrillar collagens, variable regions alternative splicing, intrinsic disorder, and tyrosine sulfation. Protein Cell 3:419–433
26. Alcaide-Ruggiero L, Molina-Hernandez V, Granados MM et al (2021) Main and minor types of collagens in the articular cartilage: the role of collagens in repair tissue evaluation in chondral defects. Int J Mol Sci 22:13329
27. Parsons P, Gilbert SJ, Vaughan-Thomas A et al (2011) Type IX collagen interacts with fibronectin providing an important molecular bridge in articular cartilage. J Biol Chem 286:34986–34997
28. Fung A, Sun M, Soslowsky LJ et al (2022) Targeted conditional collagen XII deletion alters tendon function. Matrix Biol Plus 16:100123
29. Bai X, Dilworth DJ, Weng YC et al (2009) Developmental distribution of collagen IV isoforms and relevance to ocular diseases. Matrix Biol 28:194–201
30. Wu Y, Ge G (2019) Complexity of type IV collagens: from network assembly to function. Biol Chem 400:565–574
31. Timpl R, Risteli J, Bachinger HP (1979) Identification of a new basement membrane collagen by the aid of a large fragment resistant to bacterial collagenase. FEBS Lett 101:265–268
32. Keene DR, Sakai LY, Lunstrum GP et al (1987) Type VII collagen forms an extended network of anchoring fibrils. J Cell Biol 104:611–621
33. Lapiere JC, Chen JD, Iwasaki T et al (1994) Type VII collagen specifically binds fibronectin via a unique subdomain within the collagenous triple helix. J Invest Dermatol 103:637–641
34. Zimmer KP, Schumann H, Mecklenbeck S et al (2002) Esophageal stenosis in childhood: dystrophic epidermolysis bullosa without skin blistering due to collagen VII mutations. Gastroenterology 122:220–225
35. Gara SK, Grumati P, Urciuolo A et al (2008) Three novel collagen VI chains with high homology to the alpha3 chain. J Biol Chem 283:10658–10670
36. Cescon M, Gattazzo F, Chen P et al (2015) Collagen VI at a glance. J Cell Sci 128:3525–3531
37. Scacheri PC, Gillanders EM, Subramony SH et al (2002) Novel mutations in collagen VI genes: expansion of the Bethlem myopathy phenotype. Neurology 58:593–602

38. Oh SP, Kamagata Y, Muragaki Y et al (1994) Isolation and sequencing of cDNAs for proteins with multiple domains of Gly-Xaa-Yaa repeats identify a distinct family of collagenous proteins. Proc Natl Acad Sci U S A 91:4229–4233
39. Riso E-M, Kaasik P, Seene T (2016) Remodelling of skeletal muscle extracellular matrix: effect of unloading and reloading. In: Composition and function of the extracellular matrix in the human body, pp 45–68
40. Kawashima H, Watanabe N, Hirose M et al (2003) Collagen XVIII, a basement membrane heparan sulfate proteoglycan, interacts with L-selectin and monocyte chemoattractant protein-1. J Biol Chem 278:13069–13076
41. Franzke CW, Bruckner P, Bruckner-Tuderman L (2005) Collagenous transmembrane proteins: recent insights into biology and pathology. J Biol Chem 280:4005–4008
42. Latvanlehto A, Snellman A, Tu H et al (2003) Type XIII collagen and some other transmembrane collagens contain two separate coiled-coil motifs, which may function as independent oligomerization domains. J Biol Chem 278:37590–37599
43. Nishie W, Kiritsi D, Nystrom A et al (2011) Dynamic interactions of epidermal collagen XVII with the extracellular matrix: laminin 332 as a major binding partner. Am J Pathol 179:829–837
44. Banyard J, Bao L, Zetter BR (2003) Type XXIII collagen, a new transmembrane collagen identified in metastatic tumor cells. J Biol Chem 278:20989–20994
45. Tanaka T, Wakabayashi T, Oizumi H et al (2014) CLAC-P/collagen type XXV is required for the intramuscular innervation of motoneurons during neuromuscular development. J Neurosci 34:1370–1379
46. Sato K, Yomogida K, Wada T et al (2002) Type XXVI collagen, a new member of the collagen family, is specifically expressed in the testis and ovary. J Biol Chem 277:37678–37684
47. Kular JK, Basu S, Sharma RI (2014) The extracellular matrix: Structure, composition, age-related differences, tools for analysis and applications for tissue engineering. J Tissue Eng 5:2041731414557112
48. Zeltz C, Gullberg D (2016) The integrin-collagen connection—a glue for tissue repair? J Cell Sci 129:653–664
49. Leitinger B, Hohenester E (2007) Mammalian collagen receptors. Matrix Biol 26:146–155
50. Multhaupt HA, Leitinger B, Gullberg D et al (2016) Extracellular matrix component signaling in cancer. Adv Drug Deliv Rev 97:28–40
51. Casal JI, Bartolome RA (2018) RGD cadherins and alpha2beta1 integrin in cancer metastasis: a dangerous liaison. Biochim Biophys Acta Rev Cancer 1869:321–332
52. Jokinen J, Dadu E, Nykvist P et al (2004) Integrin-mediated cell adhesion to type I collagen fibrils. J Biol Chem 279:31956–31963
53. Knight CG, Morton LF, Peachey AR et al (2000) The collagen-binding A-domains of integrins alpha(1)beta(1) and alpha(2)beta(1) recognize the same specific amino acid sequence, GFOGER, in native (triple-helical) collagens. J Biol Chem 275:35–40
54. Tuckwell D, Humphries M (1996) Integrin-collagen binding. Semin Cell DEV Biol 7:649–657
55. Pawelec KM, Best SM, Cameron RE (2016) Collagen: a network for regenerative medicine. J Mater Chem B 4:6484–6496
56. Zhang WM, Kapyla J, Puranen JS et al (2003) Alpha 11beta 1 integrin recognizes the GFOGER sequence in interstitial collagens. J Biol Chem 278:7270–7277
57. Konitsiotis AD, Raynal N, Bihan D et al (2008) Characterization of high affinity binding motifs for the discoidin domain receptor DDR2 in collagen. J Biol Chem 283:6861–6868
58. Dai SY, Chalmers MJ, Bruning J et al (2008) Prediction of the tissue-specificity of selective estrogen receptor modulators by using a single biochemical method. P Natl Acad Sci USA 105:7171–7176
59. Carafoli F, Bihan D, Stathopoulos S et al (2009) Crystallographic insight into collagen recognition by discoidin domain receptor 2. Structure 17:1573–1581
60. Borst O, Gawaz M (2021) Glycoprotein VI—novel target in antiplatelet medication. Pharmacol Ther 217:107630
61. Herr AB, Farndale RW (2009) Structural insights into the interactions between platelet receptors and Fibrillar Collagen. J Biol Chem 284:19781–19785

62. Rayes J, Watson SP, Nieswandt B (2019) Functional significance of the platelet immune receptors GPVI and CLEC-2. J Clin Invest 129:12–23
63. Smethurst PA, Onley DJ, Jarvis GE et al (2007) Structural basis for the platelet-collagen interaction—the smallest motif within collagen that recognizes and activates platelet glycoprotein VI contains two glycine-proline-hydroxyproline triplets. J Biol Chem 282:1296–1304
64. Jarvis GE, Raynal N, Langford JP et al (2008) Identification of a major GpVI-binding locus in human type III collagen. Blood 111:4986–4996
65. Tourdot BE, Brenner MK, Keough KC et al (2013) Immunoreceptor tyrosine-based inhibitory motif (ITIM)-mediated inhibitory signaling is regulated by sequential phosphorylation mediated by distinct nonreceptor tyrosine kinases: a case study involving PECAM-1. Biochemistry 52:2597–2608
66. Farndale RW, Sixma JJ, Barnes MJ (2004) The role of collagen in thrombosis and hemostasis. J Thromb Haemost 2:561–573
67. Morton LF, Hargreves PG, Farndale RW (1995) Integrin a2bl-independent activation of platelets by simple collagen-like peptides: collagen tertiary (triple-helical) and quaternary (polymeric) structures are sufficient alone for a2bl -independent platelet reactivity. Biochem J 306:337–344
68. Meyaard L, Hurenkamp J, Clevers H et al (1999) Leukocyte-associated Ig-Like Receptor-1 functions as an inhibitory receptor on cytotoxic T cells. J Immunol 162:5800–5804
69. Boskovic J, Arnold JN, Stilion R et al (2006) Structural model for the mannose receptor family uncovered by electron microscopy of Endo180 and the mannose receptor. J Biol Chem 281:8780–8787
70. Prince ME, Sivanandan R, Kaczorowski A et al (2007) Identification of a subpopulation of cells with cancer stem cell properties in head and neck squamous cell carcinoma. Proc Natl Acad Sci U S A 104:973–978
71. Wienke D, MacFadyen JR, Isacke CM (2003) Identification and characterization of the endocytic transmembrane glycoprotein Endo180 as a novel collagen receptor. Mol Biol Cell 14:3592–3604
72. Bhowmick M, Tokmina-Roszyk D, Onwuha-Ekpete L et al (2017) Second generation triple-helical peptide inhibitors of matrix metalloproteinases. J Med Chem 60:3814–3827
73. Knapinska AM, Hart M, Drotleff G et al (2019) Matrix metalloproteinase triple-helical peptide inhibitors: potential cross-reactivity with Caspase-11. Molecules 24:4355
74. Du L, Wang H, He L (2008) CD44 is of functional importance for colorectal cancer stem cells. Clin Cancer Res 14:7964–7967
75. Lauer-Fields JL, Malkar NB, Richet G et al (2003) Melanoma cell CD44 interaction with the alpha 1(IV)1263–1277 region from basement membrane collagen is modulated by ligand glycosylation. J Biol Chem 278:14321–14330
76. Lisman T, Raynal N, Groeneveld D et al (2006) A single high-affinity binding site for von Willebrand factor in collagen III, identified using synthetic triple-helical peptides. Blood 108:3753–3756
77. De Pasquale V, Pavone LM (2020) Heparan sulfate proteoglycan signaling in tumor microenvironment. Int J Mol Sci 21:6588
78. Kuivaniemi H, Tromp G, Prockop DJ (1997) Mutations in fibrillar collagens (types I, II, III, and XI), fibril-associated collagen (type IX), and network-forming collagen (type X) cause a spectrum of diseases of bone, cartilage, and blood vessels. Hum Mutat 9:300–315
79. Gajko-Galicka A (2002) Mutations in type I collagen genes resulting in osteogenesis imperfecta in humans. Acta Biochim Pol 49:433–441
80. Myllyharju J, Kivirikko KI (2001) Collagens and collagen-related diseases. Ann Med 33(1):7–21
81. Lim J, Grafe I, Alexander S et al (2017) Genetic causes and mechanisms of Osteogenesis Imperfecta. Bone 102:40–49
82. Van Dijk FS, Sillence DO (2014) Osteogenesis imperfecta: clinical diagnosis, nomenclature and severity assessment. Am J Med Genet A 164A:1470–1481
83. Tosi LL, Floor MK, Dollar CM et al (2019) Assessing disease experience across the life span for individuals with osteogenesis imperfecta: challenges and opportunities for patient-reported outcomes (PROs) measurement: a pilot study. Orphanet J Rare Dis 14:23

84. Scheres LJJ, van Dijk FS, Harsevoort AJ et al (2018) Adults with osteogenesis imperfecta: clinical characteristics of 151 patients with a focus on bisphosphonate use and bone density measurements. Bone Rep 8:168–172
85. Sillence DO, Senn A, Danks DM (1979) Genetic heterogeneity in osteogenesis imperfecta. J Med Genet 16:101–116
86. Rauch F, Glorieux FH (2004) Osteogenesis imperfecta. Lancet 363:1377–1385
87. Forlino A, Marini JC (2016) Osteogenesis imperfecta. Lancet 387:1657–1671
88. Forlino A, Cabral WA, Barnes AM et al (2011) New perspectives on osteogenesis imperfecta. Nat Rev Endocrinol 7:540–557
89. Claeys L, Storoni S, Eekhoff M et al (2021) Collagen transport and related pathways in Osteogenesis Imperfecta. Hum Genet 140:1121–1141
90. Acke FR, Dhooge IJ, Malfait F (2012) Hearing impairment in Stickler syndrome: a systematic review. Orphanet J Rare Dis 7:84
91. Hamel BCJ, Pals G, Engels CHAM et al (1998) Ehlers-Danlos syndrome and type III collagen abnormalities: a variable clinical spectrum. Clin Genet 53:440–446
92. Byers PH (2019) Vascular Ehlers-Danlos syndrome. GeneReviews pp 1–19
93. Brady AF, Demirdas S, Fournel-Gigleux S et al (2017) The Ehlers-Danlos syndromes, rare types. Am J Med Genet C Semin Med Genet 175:70–115
94. Maziers N, Dahan K, Pirson Y (2005) From Alport syndrome to benign familial hematuria: clinical and genetic aspect. Nephrol Ther 1:90–100
95. Kashtan C (2017) Alport syndrome: facts and opinions. F1000Res 6:50
96. Heidet L, Gubler MC (2009) The renal lesions of Alport syndrome. J Am Soc Nephrol 20:1210–1215
97. Shuster S (2005) Osteoporosis, a unitary hypothesis of collagen loss in skin and bone. Med Hypotheses 65:426–432
98. Prockop DJ, Kivirikko KI (1995) Collagens: molecular biology, diseases, and potentials for therapy. Annu Rev Biochem 64:403–434

Chapter 2
Hierarchical Structure of Collagen

Abstract Collagen exhibits a complex hierarchical structure that is essential for its functionality. Its primary structure consists of a repeating (Gly-X-Y)n sequence, enriched with glycine, proline, and hydroxyproline, crucial for maintaining the stability of its triple helix conformation. The secondary structure features a left-handed polyproline II (PPII) helix, resulting from the repetitive sequence and restricted rotation around proline and hydroxyproline bonds. In its tertiary structure, collagen forms a triple helix stabilized by hydrogen bonds, hydration networks, and side-chain interactions. These triple helices further assemble into microfibrils that display a quarter-staggered arrangement and an axial D periodicity of approximately 67 nm. Across different tissues, collagen demonstrates distinct structural patterns, such as the highly organized, parallel fibrils found in tendons. These variations in collagen organization and composition highlight its crucial role in maintaining the structural integrity and functional capabilities of connective tissues.

Keywords Structural protein · Primary structure · Secondary structure · Tertiary structure · Higher-order structure

2.1 Primary Structure

The primary structure of a protein is typically defined as the amino acid sequence of its polypeptide chain, but this definition can be expanded by including modifications to amino acid side chains, cross-links and other covalent structures [1]. The amino acid sequence is determined by a structural gene, while posttranslational events are responsible for creating the covalent structure [2]. Both aspects are essential for the structure and function of a protein. In the case of collagen, its primary structure involves extensive and varied posttranslational modifications, which must be considered [3]. Examples include the conversion of proline to 4-hydroxyproline or 3-hydroxyproline, conversion of lysine to hydroxylysine, glycosylation of hydroxylysine, and oxidative deamination of lysine and hydroxylysine [4–6].

© Springer Nature Singapore Pte Ltd. 2024

J. Xiao, *Collagen Mimetic Peptides and Their Biophysical Characterization*,
Springer Series in Biomaterials Science and Engineering 20,
https://doi.org/10.1007/978-981-10-1941-8_2

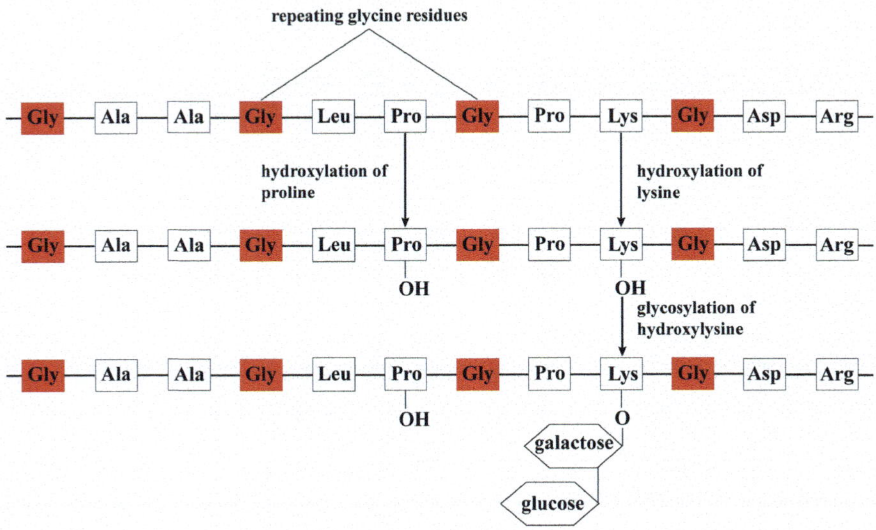

Fig. 2.1 Schematic illustration of the primary structure of collagen

Collagen has a very unique primary structure, displaying the characteristic (Gly-X-Y)$_n$ repetitive amino acid sequence pattern in its triple helix domain [7]. Among all types of natural amino acids, Gly is the smallest one with a single hydrogen atom in its side chain, and it does not have optical activity due to the presence of a non-chiral central carbon. Gly is widely viewed as the key amino acid, which is required to be present as every third residue in collagen (Fig. 2.1) [8]. It has been observed that mutations of Gly to any bulkier amino acid will break the triple helix conformation of collagen [9].

Another distinctive feature of the primary structure of collagen is a high content of proline and hydroxyproline [10, 11]. It is easily noted that the residue at the X-position can be any type, while it is frequently Pro. The residue at the Y-position is often hydroxyproline (Hyp), a derivative of Pro [12]. The preferential distribution of amino acids in the X and Y positions of the (Gly-X-Y)$_n$ sequences may provide a defining feature of collagen. Therefore, the amino acid compositions of the most abundant collagens in humans, Types I, II, and III, have been extensively analyzed to further understand the primary structure of collagen.

2.1.1 Amino Acid Distribution in Collagen Types I–III

Table 2.1 summarizes the amino acid distribution in the X and Y positions of the α_1 chain of Type I collagen. Notably, three amino acids, namely Cys, Tyr, and Trp, are completely absent, while Pro and Ala are the most abundant, constituting more than half of all amino acids in the X and Y positions. Some types of amino acids are evenly distributed among the X and Y positions, while others exhibit a clear preference for either position. For instance, Phe is exclusively present in the X-position, and Leu is predominantly restricted to this position due to severe steric hindrance in the Y-position. In contrast, Glu strongly prefers the X-position, whereas other residues, including Arg, Lys, Gln, and Thr, obviously prefer the Y position. The preferential distribution of these residues is likely to be influenced by favorable side chain-side chain or side chain-backbone interactions, rather than steric hindrance.

Tables 2.2, 2.3, and 2.4 show a similar preferential distribution of amino acids in the X and Y positions of the α_2(I), α_1(II), and α_1(III) chains of collagen. Pro and Ala are the most abundant and are relatively evenly distributed in the X and Y positions, while Cys, Tyr, and Trp are very rare. Three residues, Phe, Leu, and Glu, preferentially occupy the X-position, whereas four residues, Arg, Lys, Gln, and Thr,

Table 2.1 Amino acid distribution in the X and Y positions of the α_1 chain of Type I collagen

Residue	Number in X position	Number in Y position	Total	Percentage in X position (%)	Percentage in Y position (%)
Proline (P)	119	117	236	50.4	49.6
Glycine (G)	2	1	3	66.7	33.3
Alanine (A)	60	58	118	50.8	49.2
Isoleucine (I)	3	3	6	50.0	50.0
Leucine (L)	18	1	19	94.7	5.3
Valine (V)	9	11	20	45.0	55.0
Aspartic acid (D)	16	15	31	51.6	48.4
Glutamic acid (E)	42	5	47	89.4	10.6
Lysine (K)	12	24	36	33.3	66.6
Histidine (H)	2	0	2	100.0	0.0
Arginine (R)	9	42	51	17.6	82.4
Serine (S)	17	17	34	50.0	50.0
Threonine (T)	2	14	16	12.5	87.5
Asparagine (N)	6	5	11	54.5	45.5
Cysteine (C)	0	0	0	0.0	0.0
Glutamine (Q)	7	20	27	25.9	74.1
Phenylalanine (F)	12	0	12	100.0	0.0
Tyrosine (Y)	0	0	0	0.0	0.0
Tryptophan (W)	0	0	0	0.0	0.0
Methionine (M)	2	5	7	28.6	71.4
Total	338	338	676	50.0	50.0

consistently show a preference for the Y position. Molecular simulation studies have revealed that the terminal amino group of Gln in the Y position can form a hydrogen bond to the backbone of a nearby chain, whereas the side chain of Asn is too short to form this type of interchain hydrogen bond, regardless of whether it is located in the X or Y positions. This explains why Gln, but not Asn, exhibits a clear preference for the Y position.

One striking feature of amino acid distribution in collagen is the distinct preference for charged residues. For instance, negatively charged amino acid Glu tends to reside predominantly in the X position, while positively charged amino acids Arg and Lys more strongly favor the Y position. Both Arg and Lys contain long side chains that are capable of facilitating the formation of hydrogen bonds between their terminal amino acids and the backbone of a nearby chain if they are located in the

Table 2.2 Amino acid distribution in the X and Y positions of the α_2 chain of Type I collagen

Residue	Number in X position	Number in Y position	Total	Percentage in X position (%)	Percentage in Y position (%)
Proline (P)	109	93	202	54.0	46.0
Glycine (G)	2	1	3	66.7	33.3
Alanine (A)	48	60	108	44.4	55.6
Isoleucine (I)	10	8	18	55.6	44.4
Leucine (L)	30	3	33	90.9	9.1
Valine (V)	10	28	38	26.3	73.7
Aspartic acid (D)	8	12	20	40.0	60.0
Glutamic acid (E)	42	3	45	93.3	6.7
Lysine (K)	9	21	30	30.0	70.0
Histidine (H)	8	4	12	66.7	33.3
Arginine (R)	11	43	54	20.4	79.6
Serine (S)	16	15	31	51.6	48.4
Threonine (T)	6	13	19	31.6	68.4
Asparagine (N)	12	12	24	50.0	50.0
Cysteine (C)	0	0	0	0.0	0.0
Glutamine (Q)	5	16	21	23.8	76.2
Phenylalanine (F)	9	1	10	90.0	10.0
Tyrosine (Y)	1	0	1	100.0	0.0
Tryptophan (W)	0	0	0	0.0	0.0
Methionine (M)	1	4	5	20.0	80.0
Total	337	337	674	50.0	50.0

Table 2.3 Amino acid distribution in the X and Y positions of the α_1 chain of Type II collagen

Residue	Number in X position	Number in Y position	Total	Percentage in X position (%)	Percentage in Y position (%)
Proline (P)	111	114	225	49.3	50.7
Glycine (G)	3	1	4	75.0	25.0
Alanine (A)	57	45	102	55.9	44.1
Isoleucine (I)	7	3	10	70.0	30.0
Leucine (L)	24	1	25	96.0	4.0
Valine (V)	7	11	18	38.9	61.1
Aspartic Acid (D)	14	15	29	48.3	51.7
Glutamic acid (E)	48	5	53	90.6	9.4
Lysine (K)	13	24	37	35.1	64.9
Histidine (H)	2	0	2	100.0	0.0
Arginine (R)	9	43	52	17.3	82.7
Serine (S)	9	19	28	32.1	67.9
Threonine (T)	5	16	21	23.8	76.2
Asparagine (N)	6	6	12	50.0	50.0
Cysteine (C)	0	0	0	0.0	0.0
Glutamine (Q)	8	29	37	21.6	78.4
Phenylalanine (F)	13	0	13	100.0	0.0
Tyrosine (Y)	1	0	1	100.0	0.0
Tryptophan (W)	0	0	0	0.0	0.0
Methionine (M)	1	6	7	14.3	85.7
Total	338	338	676	50.0	50.0

Y position. Moreover, the presence of positively charged Arg and Lys in the Y position and the presence of negatively charged Glu in a nearby X position allow for favorable electrostatic interactions. Thus, the formation of potential hydrogen bonds and charged pairs could potentially serve as the determining factor for the unique distribution of charged amino acids in collagen. This phenomenon may explain why Arg and Lys tend to occupy position Y, and indirectly, why Glu is located in the X position.

Collagen's amino acid composition displays a unique preference for Thr in the Y position, as opposed to the X position [13]. Molecular simulations have shown that the OH group of Thr can form a water-bridged hydrogen bond in the Y position, but not in the X position. The presence of a Pro in the X position helps to orient the OH group by providing favorable contact with its methyl group [13]. This type of

Table 2.4 Amino acid distribution in the X and Y positions of the α_1 chain of Type III collagen

Residue	Number in X position	Number in Y position	Total	Percentage in X position (%)	Percentage in Y position (%)
Proline (P)	96	143	239	40.2	59.8
Glycine (G)	15	4	19	78.9	21.1
Alanine (A)	47	43	90	52.2	47.8
Isoleucine (I)	10	4	14	71.4	28.6
Leucine (L)	20	2	22	90.9	9.1
Valine (V)	8	3	11	72.7	27.3
Aspartic Acid (D)	10	16	26	38.5	61.5
Glutamic acid (E)	46	2	48	95.8	4.2
Lysine (K)	12	26	38	31.6	68.4
Histidine (H)	7	0	7	100.0	0.0
Arginine (R)	8	40	48	16.7	83.3
Serine (S)	27	15	42	64.3	35.7
Threonine (T)	6	9	15	40.0	60.0
Cysteine (C)	0	1	1	0.0	100.0
Asparagine (N)	7	16	23	30.4	69.6
Glutamine (Q)	9	15	24	37.5	62.5
Phenylalanine (F)	8	0	8	100.0	0.0
Tyrosine (Y)	2	0	2	100.0	0.0
Tryptophan (W)	0	0	0	0.0	0.0
Methionine (M)	5	4	9	55.6	44.4
Total	343	343	686	50.0	50.0

water-bridged hydrogen bonding is also observed for Hyp in the Y position, which plays a crucial role in stabilizing the triple helix structure of collagen [14].

2.1.2 *Overall Comparison of Amino Acid Distribution in Collagen Types I–III*

The overall picture of amino acid compositions in the X position of the α_1(I), α_2(I), α_1(II), and α_1(III) chains of interstitial collagens is illustrated in Fig. 2.2. Pro is the most populated amino acid in all collagens I–III, but its specific level varies among chains. It shows the highest percentage in the α_1(I) chain and the lowest in the α_1(III)

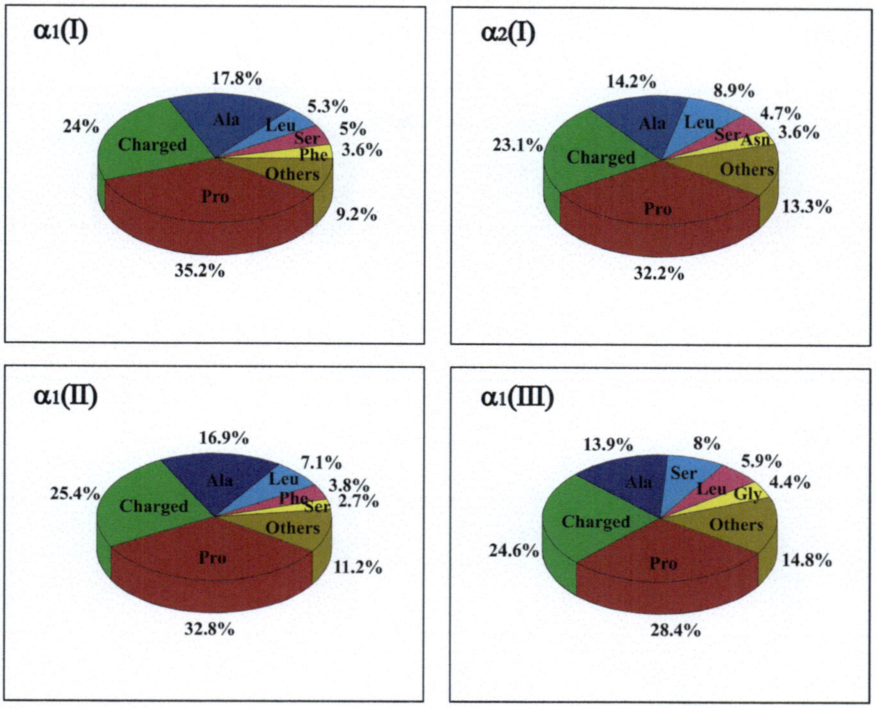

Fig. 2.2 Amino acid distribution in the X position of the α_1(I), α_2(I), α_1(II), and α_1(III) chains

chain. Charged amino acids and Ala are the next most prevalent types of amino acids. The α_1(III) chain contains significantly more Gly than other chains, suggesting that the α_1(III) chain may be the most flexible.

The overall view of amino acid compositions in the Y position is summarized in Fig. 2.3. Hyp is the most abundant amino acid in all collagen types I–III chains, with the α_2(I) chain having the lowest Hyp content among all chains. Charged amino acids and Ala are also frequently found in this position. Interestingly, significant differences exist in the distribution of other amino acids, with Thr being highly abundant in the α_1(I) and α_1(II) chains, and Val and Asn being highly abundant in the α_2(I) and α_1(III) chains, respectively. These differences in amino acid composition may contribute to the distinct biophysical properties of collagen I, II, and III.

2.1.3 The Key Role of Hydroxyproline and Hydroxylysine

4-hydroxyproline is only found in the Y, but not the X position in the triple helix domain of collagen [15]. However, the hydroxylation of proline in the Y position is not always complete, and the degree of hydroxylation may vary at a given site

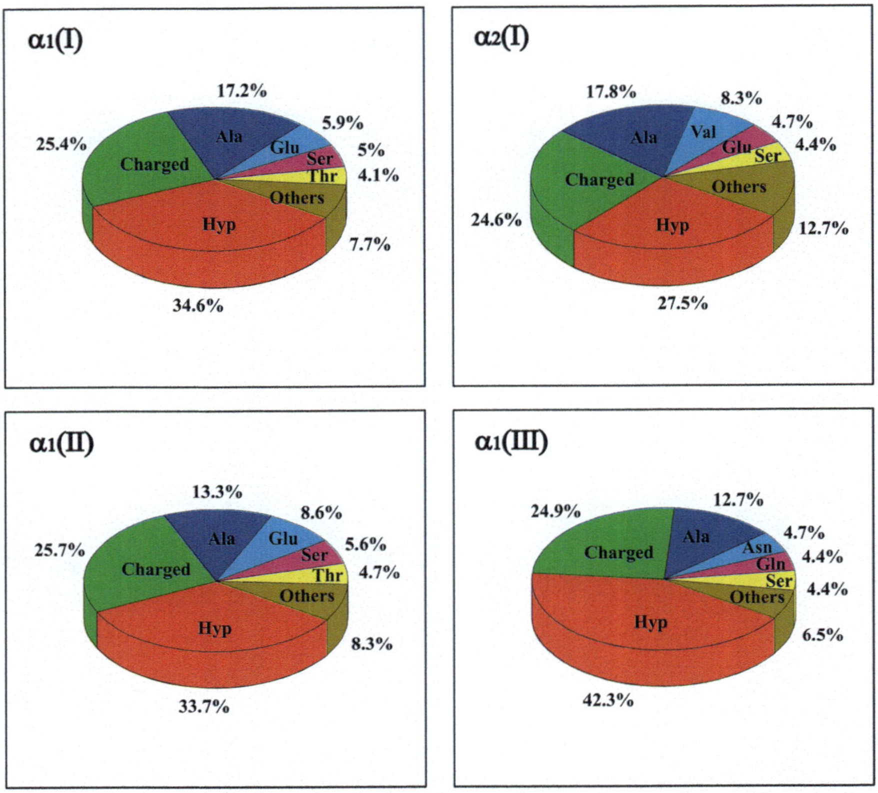

Fig. 2.3 Amino acid distribution in the Y position of the α_1(I), α_2(I), α_1(II), and α_1(III) chains

in different tissues of the same collagen. For instance, proline at position 76 is not hydroxylated in the α_1(I) chain, but it is hydroxylated in the corresponding α_1(II) chain. It has been suggested that there is a correlation between the degree of hydroxylation and the stability of the triple helix structure [6, 16]. To ensure the stability of the triple helix, a sufficient amount of hydroxyproline is necessary, and hydroxylation is halted once this threshold is reached. Additionally, 3-hydroxyproline is observed for some X positions in type IV collagen [17].

In addition to Hyp, collagen also contains another derivative amino acid, hydroxylysine, which is derived from lysine and is not directly inserted during translation [18, 19]. Depending on the types of collagens, a variety of hydroxylysines are glycosylated [6]. Enzymatic hydroxylation of proline residues occurs after polypeptide chain synthesis, specifically at the Y positions, and the conversion of lysine to hydroxylysine also only occurs at the Y positions, although to a lesser extent [19].

Compared to hydroxyproline, hydroxylysine displays much greater variability in its distribution among tissues [20]. While hydroxylysine is abundant in Type II collagen, it is not as prevalent in Type I collagen. Hydroxylysine serves as a site for covalent carbohydrate modifications, with galactosyl hydroxylysine and glucosyl galactosyl hydroxylysine being identified as carbohydrate constituents in Type I and II collagen, reaching approximately 10% in Type II collagen. In Type I collagen, only a few hydroxylysine sites have been identified for glycosylation, whereas in Type II collagen, a much larger number of hydroxylysine sites become fully glycosylated [21].

The cross-linking of lysine residues in the terminal nonhelical region is well known to be critical for the assembly of collagen into fibrils [6, 22]. Interestingly, two remarkably similar regions have been located in the triple helix domain of collagen, and they share a common sequence GMKGHRGFSGL. Both sites contain very rare residues of hydroxylysine and histidine, which may provide cross-linking sites in the triple helix domain of collagen [23].

2.2 Secondary Structure

Collagen is unique among proteins due to its distinctive left-handed polyproline II (PPII) conformation in each of its three α chains, unlike globular proteins with dominant α-helix and β-sheet secondary structures (Fig. 2.4) [4, 6]. The amino acid sequence of each α chain of collagen displays the repetitive $(\text{Gly-X-Y})_n$ pattern, with Pro and Hyp frequently occupying the X and Y positions, respectively (Fig. 2.4A). The pyrrolidine ring of Pro and Hyp restricts rotation of the bond between their C_α and the amide nitrogen, which greatly enhances the formation of the left-handed PPII conformation of collagen. The left-handed PPII conformation of collagen is relatively stretched with a pitch of 0.86 nm, which is remarkably larger than that of the α-helix (0.54 nm). The per turn contains 3.33 amino acid residues, while the axial repeat is reduced to 0.29 nm per residue (Fig. 2.4B) [16].

2.3 Tertiary Structure

Collagen's characteristic triple helix conformation is a result of its unique $(\text{Gly-X-Y})_n$ sequence pattern and high content of imino acids. The triple helix is composed of three left-handed PPII-like helices that are supercoiled around a common axis in a right-handed manner (Fig. 2.5). Within the triple helix, the three monomer chains are structurally differentiated as the leading, middle, and lagging chains, with a one-residue stagger between them. The close packing of the three chains can only accommodate Gly as every third residue, resulting in the repetitive $(\text{Gly-X-Y})_n$ sequence pattern. While the X and Y positions can be almost any amino acid, they are frequently occupied by Pro and Hyp, respectively. Both Pro and Hyp are essential for triple helix stability, and the strongest nucleating sequence to promote triple helix formation is Gly-Pro-Hyp [10, 11, 24]. Gly residues are buried deeply within the triple helix, with their amide protons directed towards the center, while X and Y residues are situated on the outside and their amide protons are exposed

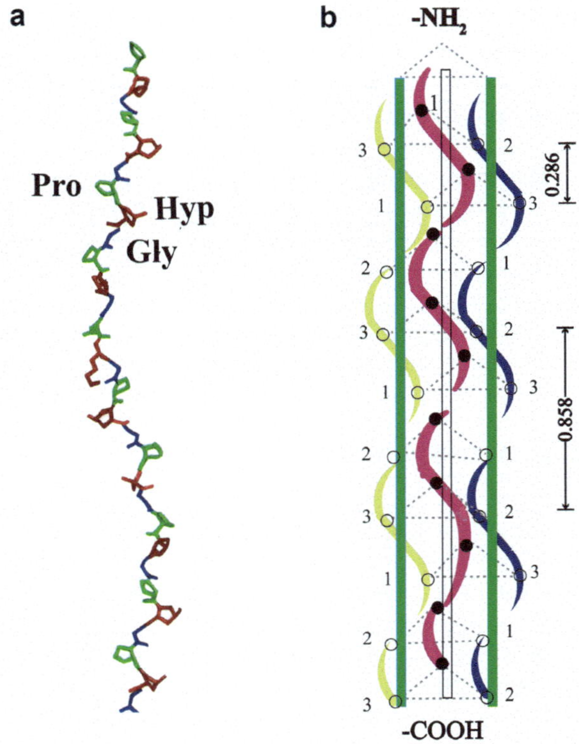

Fig. 2.4 The structure model of the left-handed polyproline II conformation of a monomer chain of collagen

to solvent. The distinct patterns in hydrogen bonding, hydration networks, and side-chain interactions provided by the $(Gly\text{-}X\text{-}Y)_n$ sequences all contribute to the defining rod-like structure of collagen.

2.3.1 Hydrogen Bonding

Hydrogen bonding is crucial in stabilizing the triple helix structure of collagen, and it occurs through a repetitive backbone hydrogen bonding network. All amino acids in the β-sheet or α-helix structure participate in the hydrogen bond formation, however, only some amino acids in the triple helix structure take part in the formation of hydrogen bonding [25]. One of the most important hydrogen bonds in collagen is formed between the NH of Gly in one chain and the $C = O$ of the residue in the X position of a nearby chain. When the residue in the X position is any amino acid other than Pro, a second interchain hydrogen bond forms between the NH group of the X-position residue and the $C = O$ of the Gly residue from a different neighboring chain (Fig. 2.6). This second set of hydrogen bonds is not directly formed, but is mediated by one water molecule and a different chain other than the one in the first set. A third

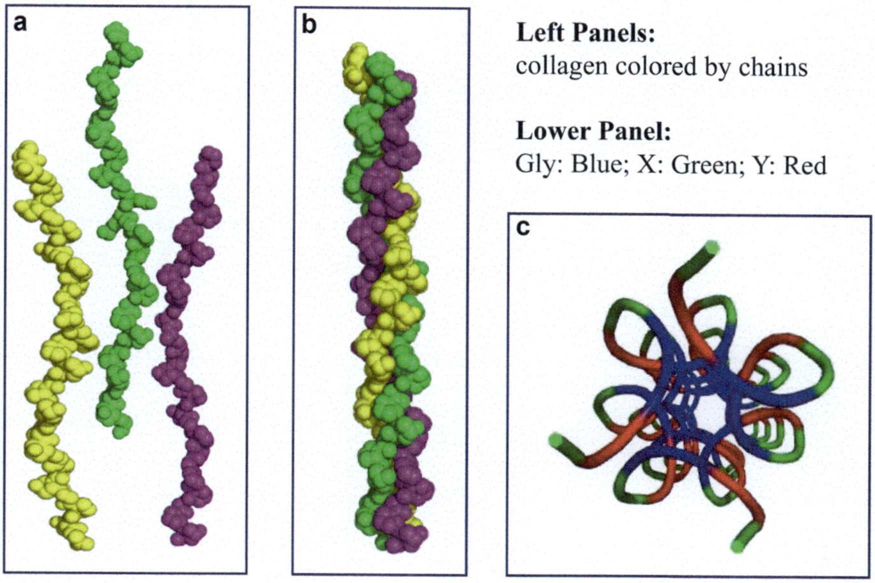

Fig. 2.5 Triple helical structure of collagen. A and B. Vertical view of the sphere model of collagen triple helical structure. Three monomer chains are right-handed supercoiled about a central axis, with one-residue staggering between different chains (the leading chain colored in yellow, the middle chain colored in green and the lagging chain colored in Fuscia). C. Cross-section view of the cartoon model of collagen triple helical structure. Gly is closely buried inside the triple helix (Gly colored in blue, X colored in green, Y colored in red). All the models are generated on the X-ray structure of peptide $(POG)_4EKG(POG)_5$ (PDB # 1QSU) by PyMol

type of hydrogen bond may exist between water molecules and a nearby Hyp or other amino acids with proper side chains. Hydrogen exchange experiments of collagen mimetic peptides demonstrate similarly slow exchanging rates for the NH groups of X-position residue as well as Gly, confirming the existence of two types of hydrogen bonding. The presence of water-bridged hydrogen bonds probably contributes to stability, and may play a key factor to enhance the triple helix regions lacking Pro.

2.3.2 Hydration Network

The presence of ordered water is an essential component of the collagen molecule, playing a vital role in preserving the molecular arrangement within collagen fibrils [15]. The collagen triple helix contains tightly bound water molecules that form a highly organized hydration network. It is believed that water binds to free backbone carbonyls as well as abundant hydroxyproline. The enhanced stabilization of collagen by hydroxyproline has been shown to contribute to the increased enthalpy, indicating the presence of a water-bridged hydrogen-bonding network. The ordered

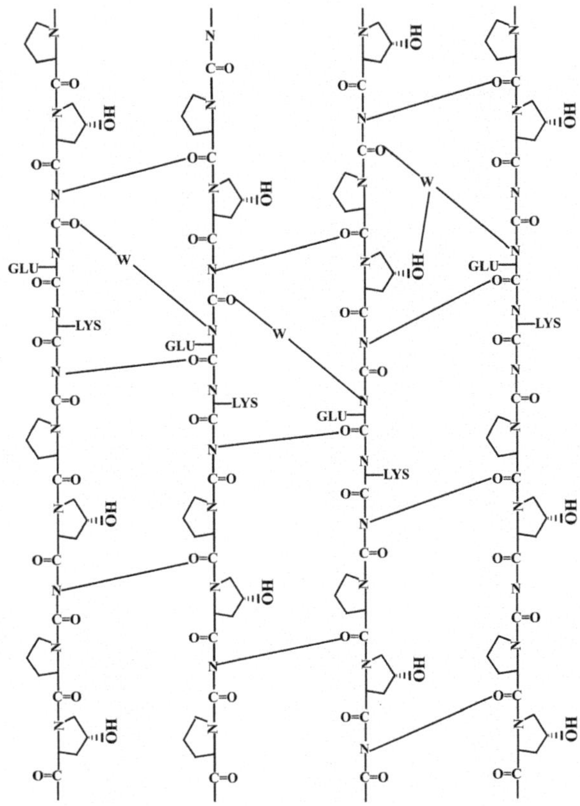

Fig. 2.6 Schematic illustration of the hydrogen bonding patterns of collagen peptide EKG (PDB 1QSU)

water network has been best visualized by X-ray structures of collagen mimetic peptides (Fig. 2.7), where water molecules are observed connecting C = O groups and Hyp hydroxyl groups within and between molecules. Notably, repetitive networks of these water patterns have been identified along the polypeptide chain. Furthermore, water molecules may prefer to cluster near hydrophobic surfaces in the triple-helical cylinder of hydration. To conclude, the extended hydration network is a prominent structural feature of collagen, as shown in high-resolution crystal structures of collagen mimetic peptides.

2.3.3 Side-Chain Interactions

The solvent-exposed side chains in the X and Y positions of the collagen triple helix provide numerous opportunities for interactions with other side chains, backbone carbonyl groups, and solvent molecules. In the crystal structures of various collagen mimetic peptides, Lys and Arg side chains in the Y position have been observed to form hydrogen bonds with the carbonyl groups of Y-position residues in adjacent

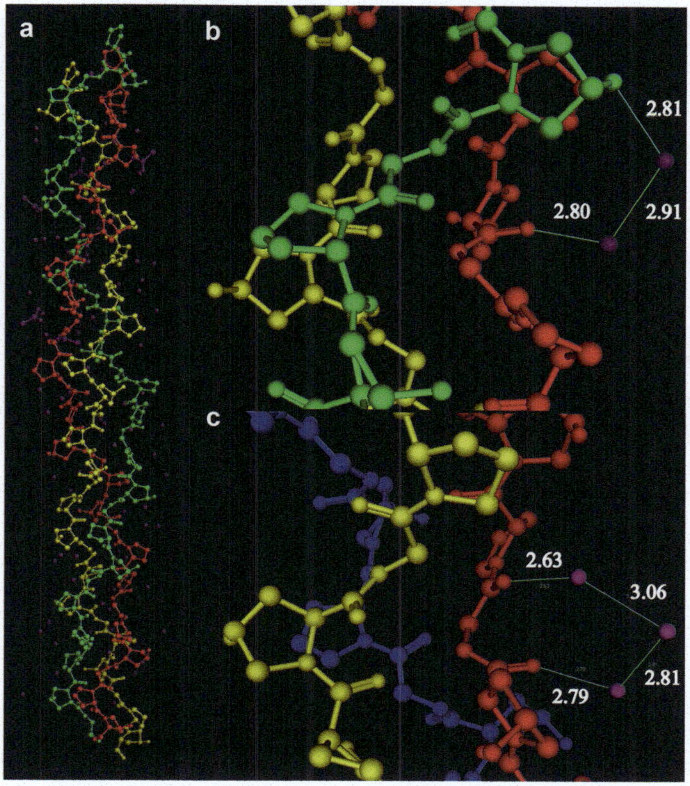

Fig. 2.7 Schematic representation of the hydration network of collagen peptide $(POG)_{10}$[G-A] (PDB 1CAG). **A** A network of water molecules was observed in close proximity to the triple helix in the high-resolution crystal structure. **B** Inter-chain water bridges involving two water molecules. **C** Inter-chain water bridges involving three water molecules. Water molecule is presented as pink balls

chains. Additionally, the Arg side chain shows hydrophobic interactions with Leu and Ile side chains from the same or neighboring chains, leading to the formation of hydrophobic clusters. The Y-position Thr side chain is observed to interact with water, mediating hydrogen bonds between the NH group of the X-position residue and the Gly carbonyl group. Taking peptide EKG (with sequence of $(Pro\text{-}Hyp\text{-}Gly)_4$-Glu-Lys-Gly-$(Pro\text{-}Hyp\text{-}Gly)_5$) as an example, interactions between the oxygen atom of Glu13 side chain and the hydroxyl group of Hyp71, as well as between the side-chain nitrogen atom of Lys44 and the Lys74 carbonyl group were observed (Fig. 2.8). These side-chain orientations and interactions seen in crystal structures highlight the significant role of side-chain contacts in maintaining the triple helix structure.

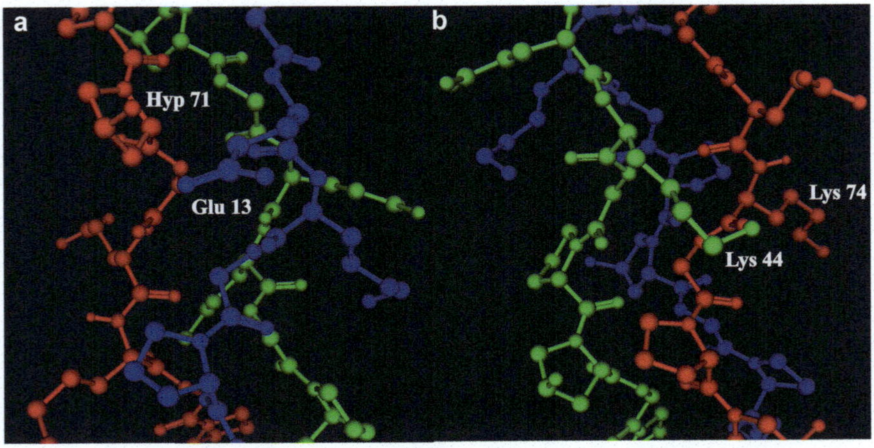

Fig. 2.8 Schematic illustration of the interactions involving the side-chain of Glu and Lys of collagen peptide EKG (PDB 1QSU)

2.4 Higher-Order Structure

Collagen, as the most abundant structural protein with excellent mechanical properties, provides one classical example of a self-assembled biological system [4, 6]. Its hierarchical structure plays a determinant role in its mechanical properties and the function of the resultant extracellular matrix. The self-assembly process of fibrous collagen can be divided into four distinct stages: (1) Modification of the α chains to form procollagen in the rough endoplasmic reticulum; (2) Packaging of procollagen in the Golgi apparatus; (3) Cleavage of the procollagen to form the mature collagen molecule in the extracellular space; (4) Fibril formation of mature collagen assisted by cross-linking. Each stage is critical for the formation of functional collagen fibers with specific mechanical properties (Fig. 2.9) [26].

During the first step, the precursor of collagen containing a signal recognition sequence is imported to the rough endoplasmic reticulum where it undergoes post-translational modifications. Lysyl and prolyl hydroxylase enzymes catalyze the hydroxylation of lysine and proline residues, respectively, and the newly formed lysine hydroxyl groups are then glycosylated via O-linked glycosylation. After these modifications, the three alpha chains self-assemble to form triple helical procollagen, which is triggered by the association of the C-propeptide domains. The high content of proline and hydroxyproline amino acids, abundant inter-chain hydrogen bonds via hydration networks, and favorable electrostatic interactions between lysine residues contribute to the stabilization of the triple helix.

In the second step, procollagen is packaged in the Golgi apparatus for export into the extracellular matrix. In the third step, the N- and C-propeptides are cleaved by proteolytic enzymes to produce mature collagen molecules with a length of 280 nm and a diameter of 1.5 nm. Collagen-specific metalloproteinases, such as

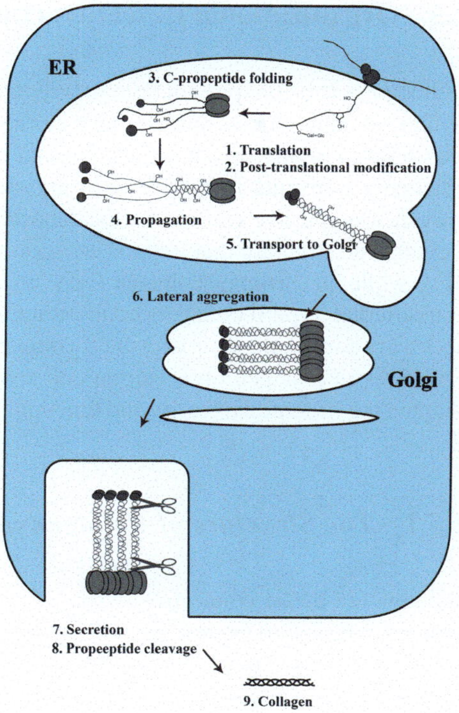

Fig. 2.9 Biosynthesis pathway of collagen

the ADAMTS (a disintegrin and metalloproteinase with thrombospondin motif), bone morphogenetic protein 1 (BMP1), and tolloid-like families, are responsible for cleaving the propeptides. The removal of the C-propeptide is a prerequisite for the initiation of fibril assembly since the C-propeptide significantly accelerates the solubility of procollagen in the extracellular matrix, thus inhibiting fibril formation. Meanwhile, the N-propeptide affects the fibril suprastructure but not fibril formation.

In the fourth step, collagen self-assembles to form microfibrils with a quarter stagger and an axial D periodicity of approximately 67 nm at the cell surface. Collagens V and XI are suggested to assist in the nucleation of microfibrils, and the fibril-associated collagens with interrupted triple helices (FACITs) integrate into the fibrils to mediate their surface properties and interactions with other extracellular matrix biomolecules. The interactions among the assembling fibrils, cell adhesion receptors, and associated macromolecules together control the supramolecular organization of the maturing fibril.

2.5 Organizational Level Structure

Collagen, a versatile protein, assumes diverse organizational structures in different tissues, playing crucial roles in their biomechanical function. In tendons, collagen manifests as highly organized and parallel fibrils, forming a robust framework that efficiently transmits forces between muscles and bones. In bone, collagen fibrils provide a supportive scaffold for the deposition of hydroxyapatite minerals, creating a composite structure that imparts exceptional hardness and resilience to the tissue. Meanwhile, in cartilage, collagen forms an intricate fibrous network predominantly composed of Type II collagen, contributing to the specific biomechanical properties necessary for cartilage to withstand compressive, shear, and tension forces. These variations in collagen organization and composition underscore its pivotal role in shaping the structural integrity and functional capabilities of these connective tissues.

2.5.1 The Structure of Collagen in Tendons

Tendons are dense connective tissues primarily composed of collagen fibers that connect muscles to bones, facilitating the transmission of forces, and promoting efficient movement and joint stability. The collagen fibrillogenesis process in tendons serves as a classical example, where bundles of aligned collagen fibrils are organized into a continuous fibrous extracellular matrix, forming a functionally mature tissue with a complex hierarchical structure (Fig. 2.10) [27]. During collagen fibrillogenesis in tendons, collagen fibril assembly initiates near the surface of tendon fibroblasts, resulting in the formation of microfibrils as the lowest order filamentous structures with a diameter of approximately 4 nm. The semi-flexible nature of microfibrils allows for their repositioning within the tendon tissue during tissue maturation and force transmission. Subsequently, lateral associations between the tapered ends of adjacent microfibrils facilitate the longitudinal growth of fibrils. Interactions between microfibrils themselves and other extracellular biomolecules, such as proteoglycans, mediate the axial growth of the microfibrils, regulating the fibril diameter. This process leads to the formation of final fibril structures within the extracellular matrix ranging in size from 10 to 500 nm. Finally, multiple fibrils are grouped together by the endotenon connective tissue, forming fibers with sizes ranging from 1 to 20 μm.

Within the tendon, fibers are bundled together to form triangular subfascicles with a diameter ranging from 15 to 400 μm. These subfascicles further assemble into larger subunits known as fascicles, which can range in size from 150 to 1000 μm. The fascicles associate with one another to form tertiary bundles, spanning from 1000 to 3000 μm, ultimately culminating in the organization of the epitenon connective tissue. Extracellular components such as glycoproteins, proteoglycans, and GAGs enhance the stability and mechanical properties of connective tissue, while collagen cross-linking provides additional reinforcement, resulting in improved mechanical stability and increased strength. The complex organization of fibrillar collagen is regulated

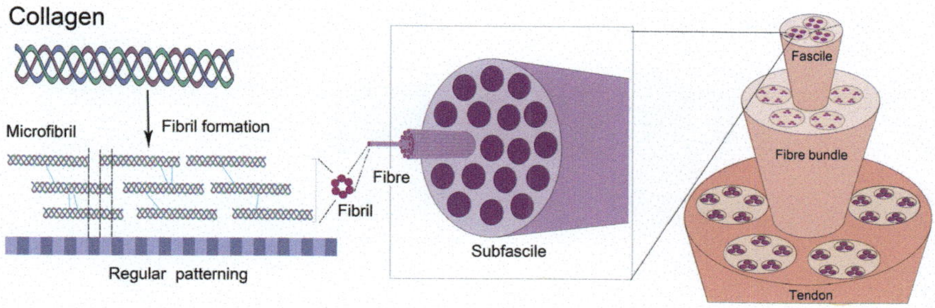

Fig. 2.10 Schematic representation of collagen fibrillogenesis in tendons

by the interplay of fibrillar and FACIT collagens, proteoglycans, and crosslinking enzymes, resulting in tissue-, stage-, and age-specific variations in fibril composition and structure. This intricate coordination generates a multi-hierarchical collagen fibrillar structure, contributing to the remarkable ability of tendon tissue to transmit forces effectively from muscle to bone [28, 29].

2.5.2 *The Structure of Collagen in Bones*

Bone and teeth are classified as hard connective tissues that undergo a fascinating biomineralization process, which ultimately yields strong and resilient structures. This process entails the deposition of calcium phosphate minerals within a collagen-rich matrix, contributing to the remarkable hardness and durability of these tissues (Fig. 2.11). X-ray diffraction studies of bone have revealed that mineralized tissues are organic–inorganic hybrid materials composed primarily of collagen and crystalline apatite [30, 31]. Notably, the deposition of hydroxyapatite (HA) crystals within the collagen matrix follows a highly organized assembly, where the crystallographic c-axis of hydroxyapatite aligns parallel to the longitudinal axis of the collagen fibril. This precise arrangement significantly enhances the strength and functional properties of bone.

Collagen serves as an active scaffold for the formation of oriented hydroxyapatite platelets during mineralization. The hydroxyapatite mineral has a propensity to nucleate preferentially within the gap regions of collagen fibrils that contain charged amino acids. Initially, calcium phosphate forms pre-nucleation clusters with loosely packed structures and negatively charged amino acids. Subsequently, the majority of hydroxyapatite crystals arrange themselves in parallel arrays. Although mineral precipitation primarily occurs in the gap zones of collagen fibrils, the crystals can grow beyond the size of the gap zone, leading to the distribution of minerals both within and outside the collagen fibrils. Notably, when exposed to calcium and phosphate-containing solutions, minerals selectively redeposit into the already

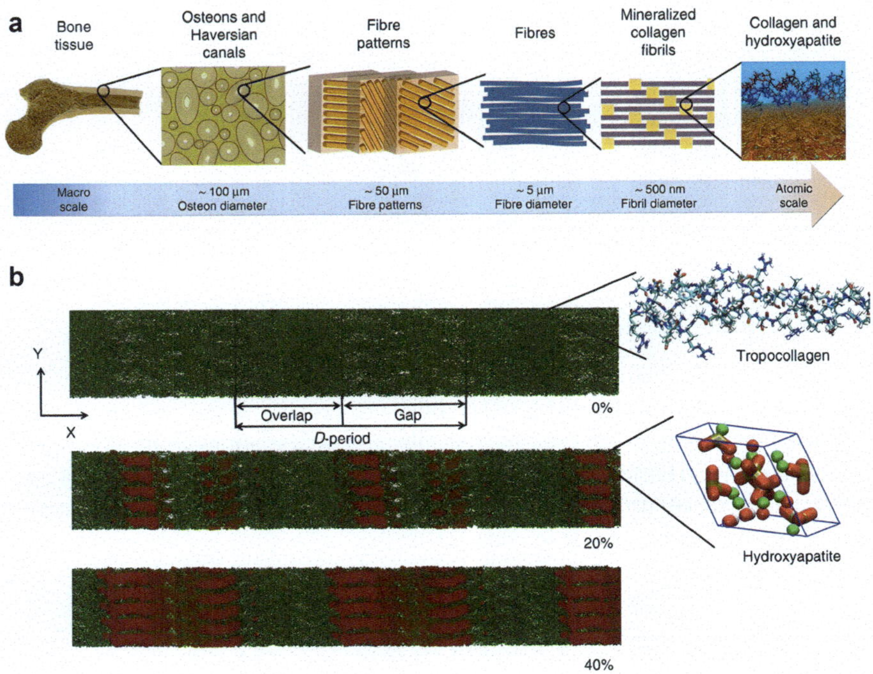

Fig. 2.11 Model structure of mineralized collagen fibrils in bone tissues. **A** Hierarchical structure of bone ranging from the macroscale skeleton to nanoscale collagen. **B** Model structures of collagen fibrils with 0, 20 and 40% mineral contents, respectively. The c-axis of the HAP (Hydroxyapatite) crystals are aligned with the fibril axis

mineralized tissues, indicating that the extracellular matrix possesses the necessary information to facilitate tissue-level mineralization.

2.5.3 The Structure of Collagen in Cartilage

Fifteen types of collagens have been identified in cartilage, including Types I, II, III, IV, V, VI, IX, X, XI, XII, XIII, XIV, XVI, XXII, and XXVII collagen, with Types I and II being the major collagens present (Fig. 2.12) [32]. Type I collagen predominates in the knee meniscus, accounting for 90% of its composition [33], while Type II collagen makes up approximately 90% of mature articular cartilage [34]. Type III collagen is crosslinked to Type II collagen and overlays Type II–IX–XI collagen copolymers in articular cartilage (Fig. 2.12).

In hyaline cartilage, Type II collagen constitutes 90–95% of the total collagen content and forms microfibrils that envelop Type XI collagen microfibrils, leading to the formation of thin fibrils with a diameter of approximately 20 nm [35, 36]. These fibrils further organize laterally, resembling ropes threaded with thin fibrils,

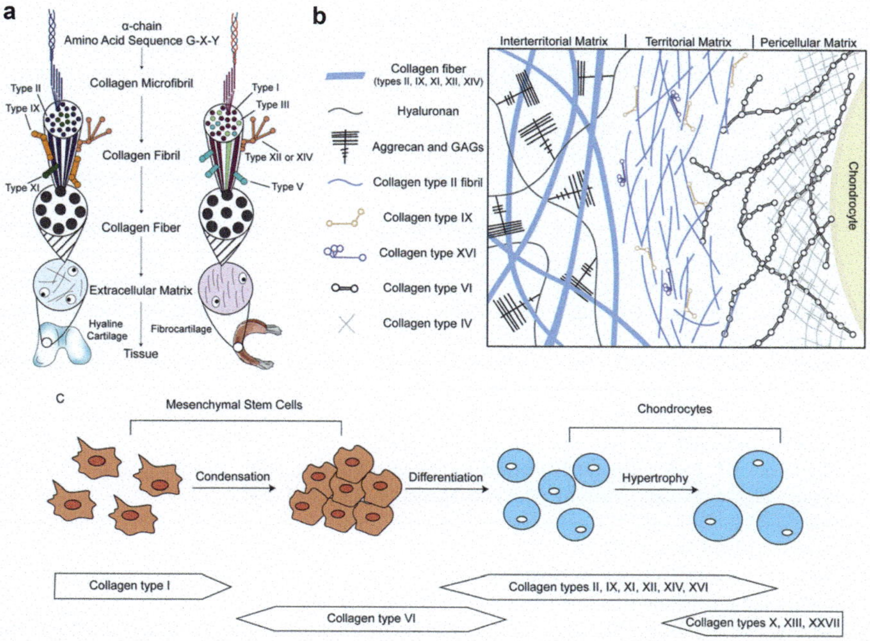

Fig. 2.12 Structure and distribution of collagen in cartilage: **A** The structure of different collagen types in cartilage ECM. **B** The distribution of different collagen types in hyaline cartilage ECM. **C** The involvement of different collagen types in the process of cartilage formation during embryonic development [32] . Reprinted from ref. [32], Copyright © 2020, with permission from Springer Nature

and form fibers with diameters ranging from 40 to 200 nm [37]. In fibrocartilage, Type I collagen exhibits the capacity for self-assembly into fibrils and can coexist with Type III and V collagen during the formation of fibrils [38]. The size of fibrils in fibrocartilage varies across different locations, ranging from 35 to 120 nm at the meniscus surface [39]. Collagen fibrils with a diameter of 100–150 nm are observed in the intervertebral disc, whereas thinner fibrils with a diameter of 20–40 nm are found in the fetal intervertebral disc [40, 41]. The unique combination of collagens and other extracellular matrix macromolecules confers specific biomechanical properties to cartilage, enabling it to withstand compression, shear, and tension forces.

References

1. Nolan C, Margoliash E (1968) Comparative aspects of primary structures of proteins. Annu Rev Biochem 37:727–790
2. Lee JM, Hammaren HM, Savitski MM et al (2023) Control of protein stability by post-translational modifications. Nat Commun 14:201–217

3. Chen X, Zhou L, Xu H et al (2020) Effect of the application of a dehydrothermal treatment on the structure and the mechanical properties of collagen film. Materials (Basel) 13:377–392
4. Ramachandran GN, Reddi AH (1976) Biochemistry of collagen. Plenum Press, New York, pp 4–545
5. Peter F (2008) Collagen: structure and mechanic. Springer, Germany
6. Brinckmann J (2005) Collagens at a glance, vol 247. Springer, Heidelberg, pp 1–6
7. Rainey JK, Goh MC (2002) A statistically derived parameterization for the collagen triple-helix. Protein Sci 11:2748–2754
8. Schroeder WA, Honnen L, Green FC (1953) Chromatographic separation and identification of some peptides in partial hydrolysates of gelatin. Proc Natl Acad Sci U S A 39:23–30
9. Sun XX, Liu Z, Zhao S et al (2019) A self-assembling collagen mimetic peptide system to simultaneously characterize the effects of osteogenesis imperfecta mutations on conformation, assembly and activity. J Mater Chem B 7:3201–3209
10. Persikov AV, Ramshaw JA, Brodsky B (2000) Collagen model peptides: sequence dependence of triple-helix stability. Biopolymers 55:436–450
11. Hyde TJ, Bryan MA, Brodsky B (2006) Sequence dependence of renucleation after a Gly mutation in model collagen peptides. J Biol Chem 281:36937–36943
12. Bhowmick M, Fields GB (2013) Stabilization of collagen-model, triple-helical peptides for in vitro and in vivo applications. Methods Mol Biol 1081:167–194
13. Salem G, Traub W (1975) Conformational implications of amino acid sequence regularities in collagen. FEBS Lett 51:94–99
14. Ramachandran GN, Bansal M, Bhatnagar RS (1973) A hypothesis on the role of hydroxyproline in stabilizing collagen structure. Biochim Biophys Acta 322:166–171
15. Bella J, Brodsky B, Berman HM (1995) Hydration structure of a collagen peptide. Structure 3:893–906
16. Engel J, Bächinger HP (2005) Collagen: primer in structure, processing and assembly. In: Structure, stability and folding of the collagen triple helix, vol 247. Springer, Heidelberg, pp 7–33
17. Pokidysheva E, Boudko S, Vranka J et al (2014) Biological role of prolyl 3-hydroxylation in type IV collagen. Proc Natl Acad Sci U S A 111:161–166
18. Yamauchi M, Sricholpech M (2012) Lysine post-translational modifications of collagen. Essays Biochem 52:113–133
19. Anazco C, Riedelsberger J, Vega-Montoto L et al (2023) Exploring the interplay between polyphenols and lysyl oxidase enzymes for maintaining extracellular matrix homeostasis. Int J Mol Sci 24:10985
20. Yamauchi M, Terajima M, Shiiba M (2019) Lysine hydroxylation and cross-linking of collagen. Methods Mol Biol 1934:309–324
21. Gerriets JE, Curwin SL, Last JA (1993) Tendon hypertrophy is associated with increased hydroxylation of nonhelical lysine residues at two specific cross-linking sites in type I collagen. J Biol Chem 268:25553–25560
22. Kivirikko KI, Pihlajaniemi T (1998) Collagen hydroxylases and the protein disulfide isomerase subunit of prolyl 4-hydroxylases. Adv Enzymol Relat Areas Mol Biol 72:325–398
23. Kawahara K, Yoshida T, Maruno T et al (2020) Spatiotemporal regulation of PEDF signaling by type I collagen remodeling. Proc Natl Acad Sci U S A 117:11450–11458
24. Thiagarajan G, Li Y, Mohs A et al (2008) Common interruptions in the repeating tripeptide sequence of non-fibrillar collagens: sequence analysis and structural studies on triple-helix peptide models. J Mol Biol 376:736–748
25. Pauling L, Corey RB (1951) The structure of fibrous proteins of the collagen-gelatin group. Proc Natl Acad Sci U S A 37:272–281
26. Koide T, Nagata K (2005) Collagen biosynthesis. Top Curr Chem 247:85–114
27. Mouw JK, Ou GQ, Weaver VM (2014) Extracellular matrix assembly: a multiscale deconstruction. Nat Rev Mol Cell Biol 15:771–785
28. Liu Y, Ramanath HS, Wang DA (2008) Tendon tissue engineering using scaffold enhancing strategies. Trends Biotechnol 26:201–209

29. Fan D, Takawale A, Lee J, Kassiri Z (2012) Cardiac fibroblasts, fibrosis and extracellular matrix remodeling in heart disease. Fibrogenesis Tissue Repair 5:15–28
30. Nair AK, Gautieri A, Chang SW et al (2013) Molecular mechanics of mineralized collagen fibrils in bone. Nat Commun 4:1724–1733
31. Launey ME, Buehler MJ, Ritchie RO (2010) On the mechanistic origins of toughness in bone. Annu Rev Mater Res 40:25–53
32. Bielajew BJ, Hu JC, Athanasiou KA (2020) Collagen: quantification, biomechanics, and role of minor subtypes in cartilage. Nat Rev Mater 5:730–747
33. Eyre DR, Wu JJ (1983) Collagen of fibrocartilage: a distinctive molecular phenotype in bovine meniscus. FEBS Lett 158:265–270
34. Eyre D (2002) Collagen of articular cartilage. Arthritis Res 4:30–35
35. Sophia Fox AJ, Bedi A, Rodeo SA (2009) The basic science of articular cartilage: structure, composition, and function. Sports Health 1:461–468
36. Holmes DF, Kadler KE (2006) The 10+4 microfibril structure of thin cartilage fibrils. Proc Natl Acad Sci U S A 103:17249–17254
37. Gottardi R, Hansen U, Raiteri R et al (2016) Supramolecular organization of collagen fibrils in healthy and osteoarthritic human knee and hip joint cartilage. PLoS ONE 11:e0163552
38. Eryilmaz E, Teizer W, Hwang W (2017) In vitro analysis of the co-assembly of type-I and type-III collagen. Cell Mol Bioeng 10:41–53
39. Petersen W, Tillmann B (1998) Collagenous fibril texture of the human knee joint menisci. Anat Embryol (Berl) 197:317–324
40. Inoue H, Takeda T (1975) Three-dimensional observation of collagen framework of lumbar intervertebral discs. Acta Orthop Scand 46:949–956
41. Hickey DS, Hukins DW (1981) Collagen fibril diameters and elastic fibres in the annulus fibrosus of human fetal intervertebral disc. J Anat 133:351–357

Chapter 3
Bioinformatic Analysis of Collagen Sequences

Abstract Understanding the connection between collagen genotypes and phenotypes, particularly in Osteogenesis Imperfecta (OI) resulting from Type I collagen missense mutations, is crucial for unraveling hereditary disease mechanisms. Bioinformatic tools are utilized to analyze collagen genotypes and phenotypes, with studies employing odds ratio (OR) analysis to explore the correlation between neighboring amino acid sequences surrounding Gly mutations and the lethality of OI, alongside the development of decision tree models for predicting OI lethality. An intriguing correlation emerges between the proximity of Proline and the absence of small, flexible residues concerning OI phenotypes: OI cases tend to be lethal when Proline is near a mutation site or when small, flexible residues are lacking. The $T_m[+1]$ model serves as a reliable predictor for mutation effects by assessing the relative stability of the Gly-X-Y triplet C-terminal to the substitution, indicating a strong link between low stability of the C-terminal triplet and OI lethality, especially for Gly-Ser and Gly-Cys mutations. Comparative analysis between pathological Gly mutations and natural interruptions underscores the nuanced impact of amino acid composition on collagen's structure and function. Integrating computational analysis with clinical data enhances our comprehension of the pathophysiological basis of hereditary diseases.

Keywords Osteogenesis imperfecta · Bioinformatics · Odds ratio analysis · Decision tree modeling

3.1 Introduction to Bioinformatic Analysis of Genotype–Phenotype Relationships

Understanding the relationship between collagen genotypes and phenotypes of connective tissue disease is essential in decoding the molecular mechanisms of hereditary diseases, as well as in developing effective diagnostic strategies [1, 2]. Osteogenesis Imperfecta (OI) is the most widely studied connective tissue disorder resulting from the missense mutation of a conserved Gly in the triple helical domain of Type I

© Springer Nature Singapore Pte Ltd. 2024

J. Xiao, *Collagen Mimetic Peptides and Their Biophysical Characterization*, Springer Series in Biomaterials Science and Engineering 20,
https://doi.org/10.1007/978-981-10-1941-8_3

collagen [3–5]. The Sillence classification system categorizes the phenotypes of OI into four types [6–8]. Type I OI is the mildest form of the disorder and is characterized by symptoms such as minimal bone deformity and low-impact fractures [4, 9]. Type II OI represents the most severe form, wherein afflicted infants endure profound bone fragility, rendering them susceptible to prenatal fractures, ultimately culminating in premature demise prior to or shortly after birth [7, 10]. Type III OI displays severe, non-lethal features such as extremely short stature and multiple bone fractures [10, 11]. Type IV OI is known for its wide array of symptoms, including moderate bone deformation, and is recognized as having the most expansive phenotype among the four types of OI [6, 12].

A comprehensive understanding of OI pathogenesis relies on comprehending the sequence determinants of the reported missense mutations that affect Type I collagen [3]. More than 700 documented missense mutations have been identified in type I collagen as contributors to OI [3, 13, 14]. Among these variations, the most prevalent substitutions involve the replacement of Gly residues with Ser and Cys [4]. Additionally, substitutions such as Arg, Val, Asp, Glu, and Ala are frequently observed [3, 13–15]. Studies have revealed that the severity of the OI phenotype associated with collagen mutations follows a certain hierarchy [16]. Mutations involving Ala and Ser substitutions tend to result in milder forms of the disease [15]. Substitutions such as Cys, Arg, and Val are associated with more severe phenotypes, while Glu and Asp substitutions are linked to the most severe forms [15]. Gaining insights into the sequence determinants of OI mutations is essential for unraveling the fundamental mechanisms of OI pathogenesis and guiding the development of targeted therapeutic strategies.

The correlation between genotype and phenotype represents a statistical relationship that aids in predicting abnormalities in individuals with similar mutations, relying on observations of affected individuals and their specific genotypes [17]. This inherent connection is established by studying the nature or location of a mutation in patients. Typically, when a specific phenotype is frequently observed in association with a particular genotype, there is a higher likelihood that an unrelated individual with the same genotype will exhibit similar abnormalities as observed in the population carrying those mutations [18, 19].

Bioinformatics has emerged as a crucial component in various biological domains, facilitating genome annotation, gene and protein expression analysis, comparative genomics, and the study of biological pathways and networks [20, 21]. With the availability of genetic sequences, bioinformatic tools have become indispensable for managing and analyzing large-scale genetic data [22, 23]. The advent of protein sequencing in the 1950s rendered manual comparison of multiple protein sequences impractical, necessitating the use of computer-based bioinformatic analyses [24–26]. In understanding the intricate relationships between genotype and phenotype, several bioinformatic and biostatistical approaches prove valuable in uncovering valuable insights.

3.1.1 Odds Ratio Analysis

Odds ratio (OR) is a statistical measure used to assess the association between an exposure and an outcome [27]. It quantifies the ratio of the odds of the outcome occurring in the presence of a particular exposure to the odds of the outcome occurring in the absence of that exposure [28–30]. The OR serves as an effect size, describing the strength of association or non-independence between two binary data values. In general, an OR of 1 suggests no association, indicating that the exposure does not affect the odds of the outcome. An OR greater than 1 indicates a positive association, suggesting that the exposure is associated with higher odds of the outcome. Conversely, an OR less than 1 indicates a negative association, implying that the exposure is associated with lower odds of the outcome. Odds ratios are frequently employed as a descriptive statistic in case–control studies, particularly in the investigation of disease phenotypes and their association with exposure variables such as genotypes. They offer valuable insights into the relationship between exposure and outcome, helping to elucidate the potential impact of specific factors on disease development or other binary outcomes of interest.

Our research group has developed an approach utilizing OR analysis to examine the relationship between neighboring amino acid sequences surrounding Gly mutations and the lethality of OI [4]. The OR is determined by evaluating the relative likelihood of a particular residue type being present versus absent at a specific position and its correlation with lethal OI. Amino acids are classified into five distinct types: Pro (Pro and Hyp), Small (Gly, Ala, Cys, Ser, and Thr), Charged (His, Lys, Arg, Glu, and Asp), Hydrophobic (Val, Ile, Leu, Met, Phe, Tyr, and Trp), and Polar (Asn and Gln). To differentiate the position of a local amino acid relative to a Gly mutation, we employ the notations Pi and Pi' to represent the ith residue preceding and following the mutation, respectively (Fig. 3.1). For example, at a specific position Pi, Pro residues occur 'a' times in all lethal OI cases and 'b' times in nonlethal cases, while non-Pro residues occur 'c' and 'd' times in lethal and nonlethal cases, respectively. The odds ratio for Pro at position Pi leading to a lethal phenotype is calculated as ad/bc (Fig. 3.1). This approach enables us to assess the relationship between neighboring amino acid sequences and the lethality of OI associated with Gly mutations.

Pi+1 Pi P2 P1 P1' P2' Pi' Pi+1'

- - - X - Y- - - - - Gly - X - Y - Gly - X - Y- Gly - - - - - - X - Y - - -

Pro	**a/b**	**25/41**
non-Pro	**c/d**	**31/156**
odds ratio	**ad/bc**	**3.1**

Fig. 3.1 Illustration of calculation of the odds ratio for Pro at position Pi leading to lethal phenotype

All instances of diagnosed cases were encompassed within this study, with each patient represented only once, regardless of the outcome being lethal or nonlethal. In cases where Gly substitutions resulted in both lethal and nonlethal outcomes, the frequencies of each were separately recorded to enable a comprehensive analysis. Different substitution types for Gly, including Ala, Ser, Cys, Glu, Arg, Asp, and Val, were found to have distinct impacts on the development of lethal OI. Therefore, separate odds ratios were calculated for each substitution type. For instance, to determine the odds ratio of the Ser mutation at the P2' position, data from both lethal and nonlethal OI cases were considered (Fig. 3.1). Pro residues were observed in 25 instances in lethal OI cases compared to 41 instances in nonlethal OI cases. Non-Pro residues were present in 31 instances in lethal OI cases and 156 instances in nonlethal OI cases. The calculated odds ratio revealed a significant association between the presence of Pro at the P2' position and a higher likelihood of lethal OI when the Ser mutation is present, with a value of 3.1 and a 95% confidence interval ranging from 1.6 to 5.8.

In addition to calculating the ORs for the presence or absence of Pro at each position (P11–P11'), ORs were also computed for four other pairs of residue types: small vs. bulkier residues, charged versus. neutral residues, hydrophobic vs. nonhydrophobic residues, and polar versus. non-polar residues. This comprehensive methodology allowed the study to explore the relationship between diverse residue types and the likelihood of lethal OI. When the odds ratios are high, a strong correlation exists between the residue type at a specific position and lethal OI, while a correlation with nonlethal OI is suggested when the odds ratios are low. To ensure the statistical robustness of these findings, Fisher's exact test was utilized, taking into account the constraints imposed by the small sample size. All tests were performed using a two-tailed approach, and statistical significance was determined with a p-value threshold of 0.05.

3.1.2 Decision Tree Modeling

Decision trees have gained significant popularity as indispensable tools in data mining due to their simplicity and ease of interpretation [31]. They provide a graphical representation of rules that partition a dataset into smaller segments [32]. The initial rule divides the data into subsets, and subsequent rules further split these subsets, forming branches and nodes. The leaves of the tree represent the final groups or unsplit nodes, where the data within each leaf share similar characteristics related to the target outcome. The primary objective of a decision tree is to segregate the data into distinct and homogeneous groups, creating a model that can capture patterns in the existing data and generalize well to new data. A well-designed decision tree strikes a balance between flexibility, ensuring accurate generalization, and interpretability for specific scenarios. There are different types of decision trees available, such as regression trees for handling continuous quantitative data and classification trees for dealing with qualitative data [33].

Our research group has developed decision tree models that utilize information obtained from odds ratio analysis to predict the lethality of OI [4]. These models incorporate critical positions identified by the odds ratio analysis as input variables, with a specific focus on the P1–P5 and P1'–P5' regions due to the limited sample size. Each input variable in the model is characterized by five nominal values: Pro, Small, Hydrophobic, Polar, or Charged. The output variable of the model is defined as the presence of the lethal OI phenotype. Ten-fold cross-validation was utilized to assess the performance of the decision tree models and determine the optimal model for predicting the lethality of OI. All decision tree analyses were performed using the SAS Enterprise Miner package (SAS Institute Inc., NC).

3.2 Bioinformatic Analysis of Osteogenesis Imperfecta Mutations in Type I Collagen

3.2.1 Preparation of Osteogenesis Imperfecta Database

The native protein sequences for α_1(I) and α_2(I) were obtained from Genbank entries NP_000079.2 and NP_000080.2, respectively [34]. Amino acids were numbered following the guidelines established by the Human Genome Variation Society [35]. The determination of lethality in OI cases was based on references from the OI consortium [3] or the original publications, encompassing patients who did not survive the immediate postnatal period and cases of prenatal lethality. Precisely identifying the phenotypes in OI can be challenging due to the complex clinical attributes [7]. This bioinformatics analysis specifically focused on investigating missense mutations of Gly within the triple helix domain of Type I collagen, known to be associated with confirmed lethality in OI. The study aimed to explore the intricate relationship between genotypes and phenotypes, incorporating a total of 615 cases with known OI lethality in the α_1(I) chain and 488 cases in the α_2(I) chain.

3.2.2 Odds Analysis of Different Regions in the α_1(I) and α_2(I) Chains

Our research group analyzed 615 cases of Gly missense mutations in the α_1(I) chain and 488 cases in the α_2(I) chain [36]. In order to facilitate the analysis, the chains were partitioned into five distinct regions (designated as Region 1, 2, 3, 4, 5), each encompassing approximately 200 amino acids (Fig. 3.2A, C). Region 1 of both the α_1(I) and α_2(I) chains of collagen had predominantly non-lethal mutations, with a few exceptions for mutations of Asp, Val, Glu, Arg, and Ser in the α_1(I) chain (Fig. 3.2B) and mutations of Arg and Val in the α_2(I) chain (Fig. 3.2D). Regions 2–5 of the α_1(I) chain mostly contained lethal mutations with Arg, Glu, Asp, or Val substitutions,

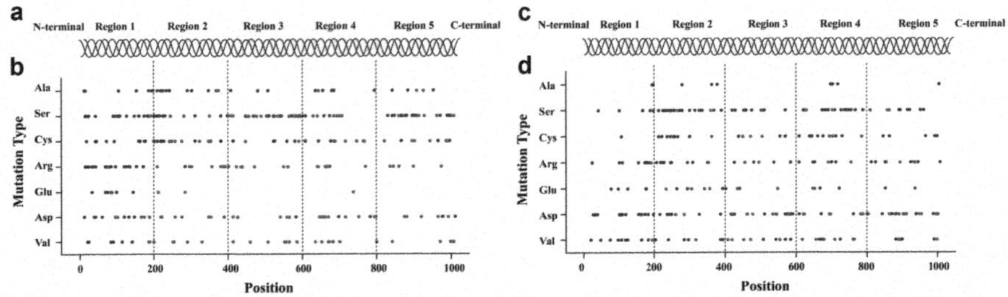

Fig. 3.2 Distribution of different types of Gly mutations (Ala, Ser, Cys, Glu, Arg, Asp and Val) along the α_1(I) and α_2(I) chains. The triple helix collagen α_1(I) (**A**) and α_2(I) (**C**) chain is divided into five regions from the N-terminal to the C-terminal: region 1, region 2, region 3, region 4 and region 5. Blue and red circles are used to represent non-lethal and lethal mutations, respectively (**B** and **D**) [10]. Reprinted from ref. [10], Copyright © 2015, with permission from Journal of Structural Biology

while non-lethal mutations were more inter-dispersed with Ser or Cys substitutions (Fig. 3.2B). In contrast, regions 2–5 of the α_2(I) chain showed more inter-dispersed non-lethal versus lethal mutations (Fig. 3.2D). Furthermore, the study revealed that different types of Gly mutations, such as Ala, Ser, Cys, Glu, Arg, Asp, and Val, had unique distribution patterns along the α_1(I) chain, indicating that both the location and type of mutation have an impact on OI lethality.

In order to determine whether the gradient model accurately depicts the correlation between sequence and OI phenotype, the likelihood of the five regions resulting in lethal versus non-lethal cases was computed using odds analysis (Fig. 3.3). The low odds observed in Region 1 of both α_1(I) and α_2(I) chains of collagen indicate that OI cases related to this region are predominantly nonlethal, aligning with previous findings (Fig. 3.3). The odds in the α_1(I) chain do not display a linear increase from the N-terminal to the C-terminal; rather, Regions 2 and 4 exhibit higher odds compared to Regions 3 and 5 (Fig. 3.3A). Similar to the α_1(I) chain, the replacement of Gly with various regions in the α_2(I) chain also exhibit varying odds of resulting in lethal OI (Fig. 3.3B). The analysis of odds indicates that the mutations observed in OI cases follow a pattern that is dependent on the specific region rather than a gradient pattern.

3.2.3 Mutation Specific Local Consensus Sequences for Lethal OI Using OR Analysis

An OR analysis was conducted to examine the potential association of a specific residue type and position near the Gly substitution in the α_1(I) chain with lethal OI. The analysis encompassed an 8-triplet region, wherein the results unveiled distinctive local consensus sequences associated with both lethal and non-lethal forms of osteogenesis imperfecta (OI) for various Gly-X mutations. (Fig. 3.4A). For Gly-Ser

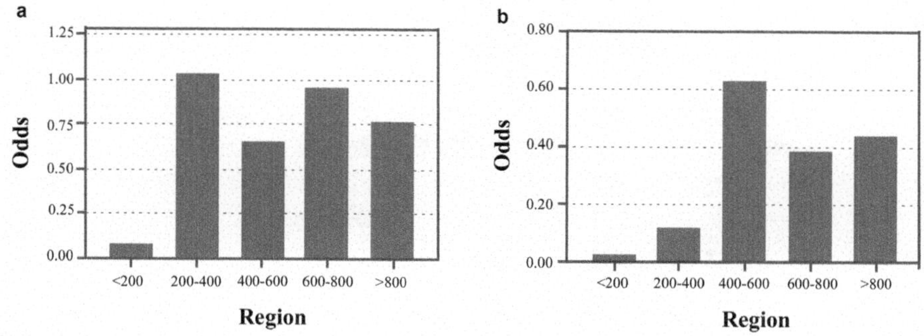

Fig. 3.3 Odds of different regions leading to lethal OI phenotypes in the α_1(I) (**A**) and α_2(I) (**B**) chains

mutations, the risk of lethal OI was significantly increased by the presence of Pro at the P1, P2, and P2' positions, while the absence of Pro at the P5' position was also found to be significantly associated with lethal OI. Similarly, for Gly-Cys mutations, the presence of Pro at the P2 and P1' positions and the absence of Pro at the P5' position were aslo correlated with lethal OI. On the other hand, for Gly to Arg mutations, the presence of Pro at the P1' position was significantly associated with lethal OI. It is noteworthy that only specific positions within the 8-triplet region, and not the entire region, were significantly linked to the OI phenotype. Furthermore, the OI lethality associated with distinct Gly-X mutations varies depending on the specific sequence environment surrounding the Gly-X position.

The effect of the identity of neighboring residues within 4-triplets of Gly substitutions with Ser, Cys or Arg on the lethality of OI was further investigated in the α_1(I) chain (Fig. 3.4B). Notably, the presence of Pro in the nearest triplet (P2–P2') was consistently found to play a pivotal role, with its presence being associated with a higher likelihood of a lethal phenotype (indicated by +). Furthermore, the presence of Pro further away from the mutation sites (P11–P7, P5'–P11') seemed to be more correlated with a non-lethal phenotype (indicated by −) (Fig. 3.4B). Additionally, the absence of small residues (P5–P2') was identified as an important factor for OI lethality. Hydrophobic residues were generally preferred at positions P2–P7' for nonlethal OI, while Arg mutations exhibited the strictest requirements. Polar residues displayed a more diverse pattern, but the presence of polar residues around Ser substitutions at positions P4 and P7' was associated with lethal OI. The presence of charged residues near the mutation, particularly at the P2 site, significantly correlated with lethal OI for Arg mutations, whereas charged residues at three sites C-terminal to the mutation were more commonly observed in nonlethal OI cases (Fig. 3.4B).

OR analysis was further performed on an 8-triplet region surrounding a Gly substitution in the α_2(I) chain to evaluate the relationship between specific residue types, their positions relative to the mutation, and lethal OI (Fig. 3.5A). The findings demonstrated that the presence of Pro and small residues at these positions exerted the most significant influence on OI lethality (Fig. 3.5A). Notably, small flexible residues

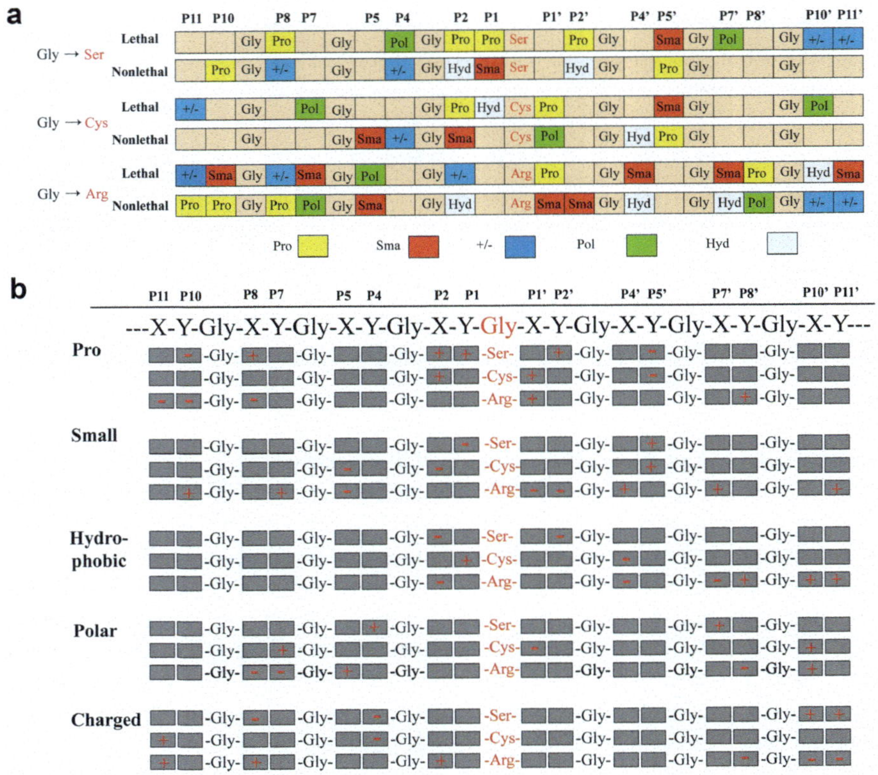

Fig. 3.4 **A** Consensus sequence patterns for Gly to Ser, Cys and Arg mutations in the α_1(I) chain that are correlated with lethal and nonlethal OI. **B** Dependency of OI lethality on the type and position of neighboring residues within 4-triplets of Gly substitutions by Ser, Cys or Arg based on odds ratio analysis in the α_1(I) chain

at the P2–P2' positions were frequently associated with nonlethal OI, while small residues located further away from the mutation site (P11–P7, P5'–P11') tended to be linked to a lethal phenotype (Fig. 3.5B). The impact of Pro residues in the α_2(I) chain displayed a complex pattern. For Ser mutations, Pro at positions P1' and P2' were associated with lethal OI, whereas Pro at positions P1 and P2 were more commonly related to nonlethal OI. Conversely, the presence of Pro at the P2' position for Arg and Asp mutations tended to result in nonlethal OI. The preference for small amino acids near Gly mutations generally contributed to nonlethal OI, while the impact of Pro residues varied depending on their specific positions within the α_2(I) chain and the type of mutation (Fig. 3.5B).

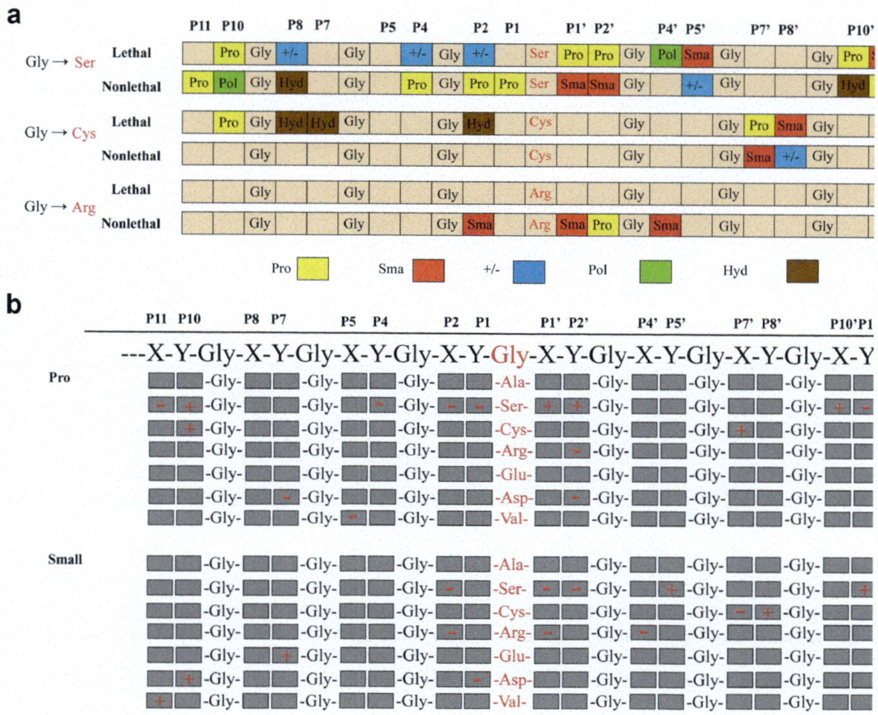

Fig. 3.5 **A** Consensus sequence patterns for Gly to Ser, Cys and Arg mutations in the α₂(I) chain that are correlated with lethal and nonlethal OI. **B** Dependency of OI lethality on the type and position of neighboring residues within 4-triplets of Gly substitutions based on odds ratio analysis in the α₂(I) chain

3.2.4 Predictive Modeling for Gly-Ser Mutations in the $\alpha_1(I)$ and $\alpha_2(I)$ Chains

Decision tree analysis was employed to develop predictive models for OI phenotypes and genotypes in the α_1(I) chain, specifically focusing on Gly to Ser mutations. Through OR analysis, several critical sites, including P4, P2, P1, P2', and P5', were identified as influential in OI. These sites were integrated into the decision tree modeling process, with P4 identified as the most significant node (Fig. 3.6).

The decision tree analysis revealed that the lethality of OI was determined by the residue type at the P1 position when the P4 residue was Pro, small, or charged. In cases where the P1 residue was simultaneously Pro, small, or charged, a nonlethal OI phenotype was generally observed. However, if the P1 residue was not Pro, small, or charged, the OI phenotype depended on the residue type at the P2 position. Charged residues at P2 were associated with a higher likelihood of lethal OI, while non-charged residues led to the OI phenotype being determined by the residue type at the P2' position. These findings highlight the significance of amino acid types at

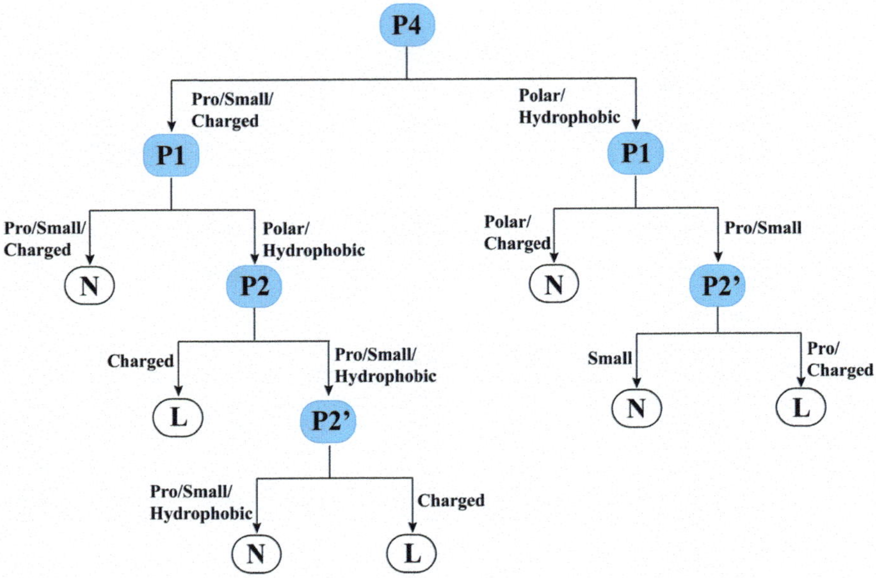

Fig. 3.6 Decision tree model of Ser substitution leading to lethal OI in the α_1(I) chain. L and N in the circles represent Lethal and Nonlethal OI cases, respectively

positions P4, P2, P1, and P2' in predicting the lethality of OI in Gly to Ser mutations in the α_1(I) chain.

The decision tree model achieved a high accuracy rate of approximately 98% in predicting nonlethal mutations for Ser substitutions in the α_1(I) chain (Table 3.1). However, the prediction of lethal mutations for Ser substitutions was more challenging, with accuracies of only 39%. Nevertheless, the decision tree model demonstrated an overall accuracy of 85% for Ser substituting residues, highlighting its effectiveness in capturing a substantial amount of information despite the difficulties in predicting lethal mutations (Table 3.1).

Moreover, a decision tree model was established specifically for Gly to Ser mutations in the α_2(I) chain, with key nodes identified at positions P4', P5', P1', and P2 (Fig. 3.7). The presence of a polar residue at the P4' site was strongly associated with a nonlethal OI phenotype. The decision tree analysis revealed that the lethality of OI depended on the residue type at the P5' position when the P4' residue

Table 3.1 Decision tree performance for Gly-Ser substitutions in the α_1(I) and α_2(I) chains

Mutant Chain	Lethal cases			Non-lethal cases			Overall corr. (%)
	Correct pred.	Wrong pred.	Correct pred. (%)	Correct pred.	Wrong pred.	Correct pred. (%)	
α_1(I)	22	34	39	194	3	98	85
α_2(I)	9	8	53	193	2	99	95

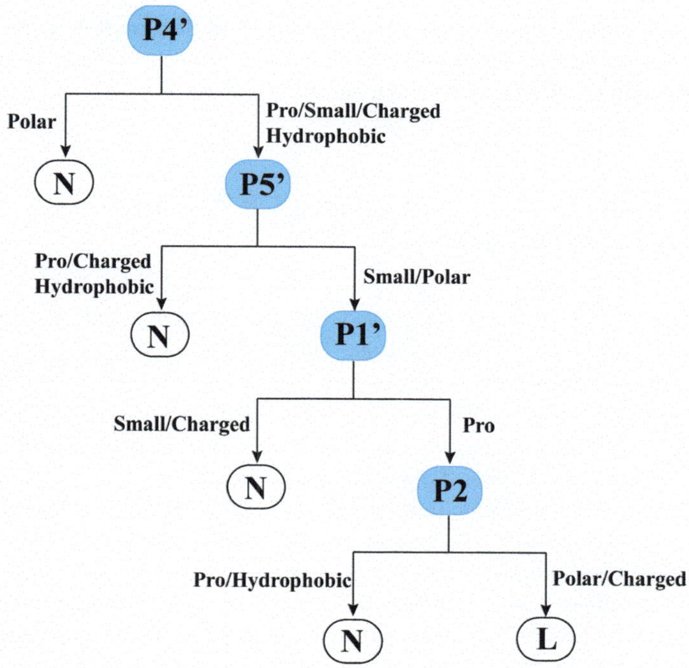

Fig. 3.7 Decision tree model of Ser substitution leading to lethal OI in the α_2(I) chain

was Pro, small, charged, or hydrophobic. Conversely, a nonlethal OI phenotype was more likely when the P5' residue was Pro, charged, or hydrophobic. The decision tree model achieved a remarkable accuracy rate of approximately 99% in predicting nonlethal mutations for Ser substitutions, while correctly identifying around 53% of lethal substitutions (Table 3.1). Overall, the decision tree model exhibited an impressive accuracy of 85%, demonstrating its effectiveness in predicting the OI lethality of Gly to Ser mutations in the α_2(I) chain (Table 3.1).

3.2.5 *Model Validation Using Novel Gly Substitutions in the α_1(I) and α_2(I) Chains*

The decision tree models were assessed for their effectiveness by testing them on previously unreported mutations from the updated Collagen Mutations Database, achieving an impressive accuracy of 85% in predicting the 13 newly reported cases of Gly-Ser/Arg/Cys mutations in the α_1(I) chain (Table 3.2). Similarly, the models accurately predicted 88% of the eight newly discovered cases of Gly-Ser substitutions in the α_2(I) chain (Table 3.2). Notably, the models were able to extrapolate to predict 15 novel cases in positions that had not been observed previously in the collagen chain. Overall, these findings underscore the efficacy of the decision tree models in predicting new mutations based on local amino acid sequences, highlighting their potential to aid in the diagnosis of collagen-related disorders.

3.3 Predicting Lethality of Collagen Gly Mutation from T_m[+1] Model

Dr. Klein has developed a composite model to predict the clinical severity of Gly mutations in the α_1 chain of type I collagen. The model utilizes T_m[+1], which represents the relative thermostability of the Gly-X-Y triplet C-terminal to the substitution, to assess the lethality of Gly-Ser mutations [37]. Statistical analysis, including a t-test, revealed a significant difference in mean T_m[+1] between lethal and nonlethal mutations, with a 4.3 °C lower T_m[+1] for lethal mutations (mean T_m[+1] = 35.9 °C) compared to nonlethal mutations (mean T_m[+1] = 40.2 °C) (Table 3.3) [37]. This finding strongly supports the hypothesis that the low stability of the C-terminal triplet is highly correlated with lethality. To evaluate the model's predictive performance, an ROC curve was generated, illustrating the trade-off between sensitivity and specificity at different T_m[+1] cutoff values (Fig. 3.8A). A T_m[+1] cutoff of 37.7 °C was determined for predicting lethality, yielding an estimated accuracy of 71% for lethal mutations and 78% for nonlethal mutations (Table 3.3).

The T_m[+1] model was further employed to forecast the phenotypes of Gly-Ser mutations in the α_2 chain of type I collagen. Remarkably, lethal Gly-Ser mutations in the α_2(I) chain exhibited significantly lower average T_m[+1] values (30.9 °C) compared to nonlethal mutations (40 °C), as confirmed by multiple statistical tests (p-values < 0.05 for t-test, Wilcoxon rank sum test, logistic regression, and permutation tests). Using a T_m[+1] cutoff value of 32.9 °C (Fig. 3.8B), the estimated prediction accuracies for lethal mutations, nonlethal mutations, and overall accuracy were found to be 75, 86, and 85%, respectively (Table 3.3). These findings highlight the substantial potential of T_m[+1] as a robust and reliable indicator for accurately predicting the lethality of Gly-Ser mutations.

A significant correlation between T_m[+1] and lethality was also observed for Gly-Cys mutations in the α_2 chain of type I collagen (Table 3.3) [37]. The mean T_m[+1] value for lethal Gly-Cys mutations (35.8 °C) was notably lower than that for nonlethal mutations (41.9 °C), as confirmed by various statistical tests including the t-test, permutation testing, Wilcoxon rank sum test, and logistic regression. Using a T_m[+1] cutoff value of 37.7 °C (Fig. 3.10C), the estimated prediction accuracies for lethal mutations, nonlethal mutations, and overall accuracy were 80, 87, and 85% respectively (Table 3.3). These results strongly support the utility of T_m[+1] as a reliable indicator for predicting the lethality of Gly-Cys mutations.

Collagen-like peptides were synthesized to experimentally validate the T_m[+1] model and investigate the impact of Gly-Ser/Cys mutations on the local structural stability of collagen. The results revealed an intriguing asymmetry in the interaction

Table 3.2 Lethality prediction for novel Gly substitutions in the α_1(I) and α_2(I) chains

Chain	Mutation type	Mutation position	Lethality	Neighboring sequences	Prediction	Prediction accuracy
α_1(I)	Gly-Ser	37	Nonlethal	GPPGEPGEPGASGPMGPRGPPGPP	Nonlethal	Yes
		199	Nonlethal	GPQGVRGEPGPPGPAGAAGPAGNP	Nonlethal	Yes
		223	Nonlethal	GADGQPGAKGANGAPGIAGAPGFP	Nonlethal	Yes
		628	Nonlethal	GARGAPGDRGEPGPPGPAGFAGPP	Nonlethal	Yes
		655	Nonlethal	GQPGAKGEPGDAGAKGDAGPPGPA	Nonlethal	Yes
		733	Nonlethal	GPAGKEGGKGPRGETGPAGRPGEV	Nonlethal	Yes
		967	Nonlethal	GASGPAGPRGPPGSAGAPGKDGLN	Nonlethal	Yes
		1000	Nonlethal	GPRGRTGDAGPVGPPGPPGPPGPP	Nonlethal	Yes
	Gly-Arg	73	Nonlethal	GRPGERGPPGPQGARGLPGTAGLP	Lethal	No
		181	Lethal	GAVGAKGEAGPQGPRGSEGPQGVR	Lethal	Yes
		214	Nonlethal	GAAGPAGNPGADGQPGAKGANGAP	Lethal	No
		661	Lethal	GEPGDAGAKGDAGPPGPAGPAGPP	Lethal	Yes
	Gly-Cys	25	Nonlethal	GPPGAPGPQGFQGPPGEPGEPGAS	Nonlethal	Yes
α_2(I)	Gly-Ser	193	Nonlethal	GNAGPAGPAGPRGEVGLPGLSGPV	Nonlethal	Yes
		202	Nonlethal	GPRGEVGLPGLSGPVGPPGNPGAN	Nonlethal	Yes
		292	Nonlethal	GPPGPSGEEGKRGPNGEAGSAGPP	Lethal	No
		427	Nonlethal	GKNGDKGHAGLAGARGAPGPDGNN	Nonlethal	Yes
		496	Nonlethal	GLHGEFGLPGPAGPRGERGPPGES	Nonlethal	Yes
		658	Nonlethal	GAKGERGAKGPKGENGVVGPTGPV	Nonlethal	Yes
		667	Nonlethal	GPKGENGVVGPTGPVGAAGPAGPN	Nonlethal	Yes
		833	Nonlethal	GPAGKEGLRGPRGDQGPVGRTGEV	Nonlethal	Yes

Table 3.3 $T_m[+1]$ models summary

chain	substitution	mean $T_m[+1]^b$ (°C)		N		p^c	$T_m[+1]$ cutoff (°C)^d	% correct^e			% coverage^f
		lethal	nonlethal	lethal	nonlethal			lethal	nonlethal	total	
COL1A1	Ser	35.9	40.2	14	36	0.03	37.7	71	78	76	82
COL1A2	Ser	30.9	40.0	4	30	0.02	32.9	75	86	85	74
COL1A2	Cys	35.8	41.9	5	14	0.008	37.7	80	87	85	90

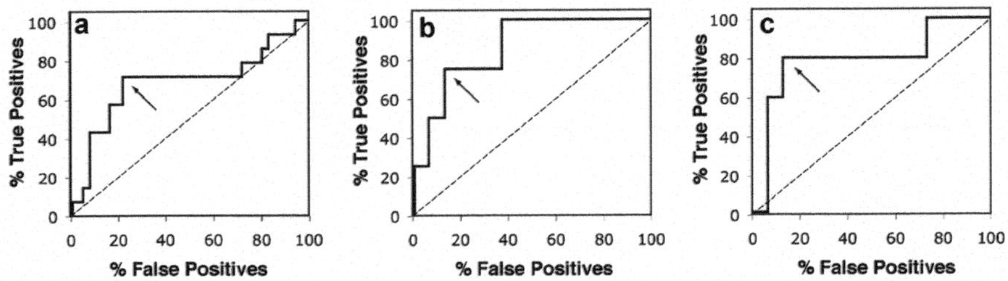

Fig. 3.8 ROC curves for Gly-Ser model along the α_1(I) chain (**A**). ROC curves for Gly-Ser model (**B**) and Gly-Cys model along the α_2(I) chain (**C**)

between a Gly-Ser mutation and its neighboring residues. Specifically, peptides with a low-stability triplet C-terminal to the mutation exhibited a significant decrease in stability, calorimetric enthalpy, and folding time compared to peptides with the low-stability triplet on the N-terminal side. This experimental evidence, combined with computational analyses, deepens our understanding of the underlying mechanisms of OI and highlights the importance of local structural stability in collagen-related disorders.

3.4 Sequential Comparison of Pathological Mutations Versus Native Interruptions

3.4.1 Differential Biological Outcomes of Gly-X Mutations and Natural Interruptions

The structural integrity of the triple helix domain in fibrillar collagens relies on the presence of Gly as every third residue, spanning approximately 1000 residues. Any substitution of a single Gly with a larger amino acid within the triple helix will initiate a disease state, such as osteogenesis imperfecta (OI), a connective tissue disorder [3, 38]. In Type I collagen alone, more than 700 distinct Gly-to-X mutations have been identified, all associated with OI [3]. In contrast, non-fibrillar collagens exhibit natural interruptions in the $(Gly\text{-}X\text{-}Y)_n$ pattern, with over 350 observed interruptions [39]. These interruptions play a crucial role in shaping the molecular structures of non-fibrillar collagens and regulating collagen binding and degradation [36]. Interestingly, while Gly substitutions and natural interruptions both disrupt the $(Gly\text{-}X\text{-}Y)_n$ sequence pattern, they lead to opposite outcomes. The underlying reasons for these contrasting effects are not yet well understood, and further research is needed to unravel the underlying mechanisms.

3.4.2 Comparison of Sequence Environments of Gly-Ala Mutations and G5G[Ala] Interruptions

In order to gain deeper insights into the divergent functional outcomes of Gly mutations and natural interruptions in collagen, our group has conducted a comparative analysis of the sequence environments of Gly-Ala mutations and G5G[Ala] interruptions [36]. A total of 33 Gly-Ala OI mutations in the α(1) chain of Type I collagen, along with 19 G5G[Ala] interruptions, were gathered from a current database, and a comparison was made between the distribution of amino acid types at each position (AA1, AA2, AA4, AA5) in the Gly-Ala mutations and G5G[Ala] interruptions (Fig. 3.9).

No notable differences were observed between Gly-Ala mutations and G5G[Ala] interruptions at positions AA1 and AA2, while a statistically significant distinction was found at position AA5 ($p = 0.03$). Moreover, both Gly-Ala mutations and G5G[Ala] interruptions displayed a pronounced prevalence of Pro residues; however, G5G[Ala] interruptions exhibited a higher proportion of hydrophobic amino acids and a reduced number of polar amino acids compared to Gly-Ala mutations. The most striking disparity was observed at position AA4, where the p-value was found to be < 0.001. Specifically, G5G[Ala] interruptions exhibited a distinctive composition characterized by 63% hydrophobic residues and 0% Pro at position AA4, clearly distinguishing them from Gly-Ala mutations. In contrast, Gly-Ala mutations at the corresponding site displayed an elevated ratio of Pro residues (39%) but a relatively lower occurrence of hydrophobic amino acids (12%). These findings indicate that variations in the amino acid composition and their subsequent impact on collagen structure and function contribute significantly to the contrasting effects of Gly mutations and natural interruptions.

3.4.3 Comparison of Sequence Environments of Gly-Ser Mutations and G5G[Ser] Interruptions

Our research group has performed an extensive investigation of Gly-Ser mutations and similar natural interruptions in collagen [40]. Two sites with the same RSE context were identified, where the Gly residue in the RGE motif is substituted with Ser. Peptide ERSEQ, with the amino acid sequence $(GPO)_6$-GER$\underline{S}$EQ-$(GPO)_6$

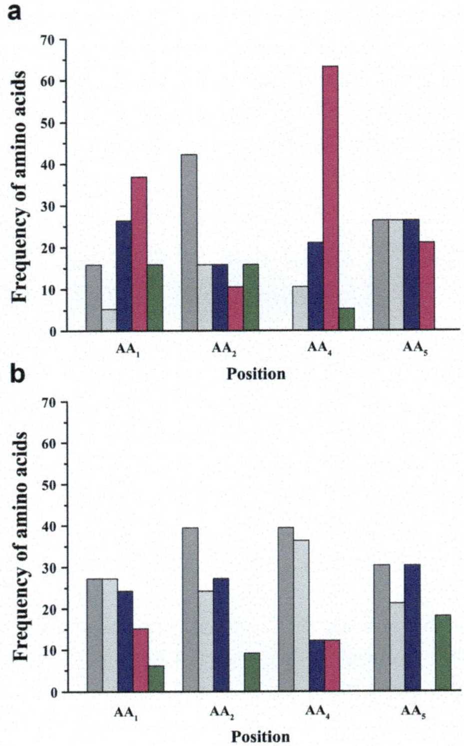

Fig. 3.9 The distribution of amino acid types in the sequence of Gly-AA_1-AA_2-Ala-AA_4-AA_5-Gly. The frequency of amino acid types in the positions (AA_1, AA_2, AA_4, AA_5) of G5G[Ala] interruptions (G5G type interruptions with a central Ala at the AA_3 site) in nonfibrillar collagens (**A**) and Gly-Ala mutations in the α_1(I) chain of Type I collagen (**B**). The amino acids are counted in five types: Pro (gray), Small (light gray), Charged (royal), Hydrophobic (pink) and Polar (olive)

(mutation site underlined), was used to model a lethal Gly-Ser mutation at site 454 in the α(1) chain of Type I collagen. Peptide DRSER with the sequence $(GPO)_6$-GDRSER-$(GPO)_6$ (interruption site underlined) was employed to model a G5G interruption at site 491–495 in the α(1) chain of Type XXIII collagen. Both ERSEQ and DRSER peptides exhibited a significant presence of charged amino acids, with the distinguishing factor being the replacement of Arg with a polar residue in DRSER.

The comparative sequence analysis revealed a significant differentiation at the fifth position (AA5) between G5G interruptions and Gly-Ser mutations, with G5G interruptions exhibiting a predominant presence of charged amino acids (~33%), contrasting with Gly-Ser mutations ($p < 0.05$) (Fig. 3.10). CD and NMR real-time folding analyses showed that peptide ERSEQ exhibited a slower folding rate compared to peptide DRSER. The two peptides serve as examples of the change of a charged Arg residue at the natural interruption site to a neutral Gln residue near a Ser mutation site in collagen sequences characterized by a highly charged context, potentially leading to diverse effects on the stability, conformation, and folding of peptides due to differences in charge, size, and side-chain interactions.

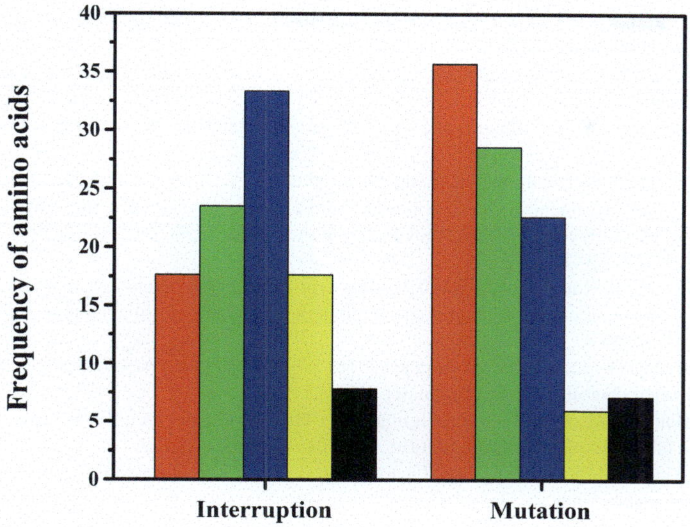

Fig. 3.10 The distribution of amino acid types at the AA_5 position in the sequence of Gly-AA_1-AA_2-AA_3-AA_4-AA_5-Gly of G5G interruptions in nonfibrillar collagens (**A**) and Gly-Ser mutations in the $\alpha(1)$ chain of Type I collagen (**B**). The amino acids are counted in five types: Pro (red), Small (green), Charged (blue), Hydrophobic (yellow) and Polar (black)

References

1. Bodian DL, Klein TE (2009) COLdb, a database linking genetic data to molecular function in fibrillar collagens. Hum Mutat 30:946–951
2. Reuter MS, Schwabe GC, Ehlers C et al (2013) Two novel distinct COL1A2 mutations highlight the complexity of genotype-phenotype correlations in osteogenesis imperfecta and related connective tissue disorders. Eur J Med Genet 56:669–673
3. Marini JC, Forlino A, Cabral WA et al (2007) Consortium for osteogenesis imperfecta mutations in the helical domain of type I collagen: regions rich in lethal mutations align with collagen binding sites for integrins and proteoglycans. Hum Mutat 28:209–221
4. Xiao J, Yang Z, Sun X et al (2015) Local amino acid sequence patterns dominate the heterogeneous phenotype for the collagen connective tissue disease Osteogenesis Imperfecta resulting from Gly mutations. J Struct Biol 192:127–137
5. Shoulders MD, Raines RT (2009) Collagen structure and stability. Annu Rev Biochem 78:929–958
6. Sillence DO, Senn A, Danks DM (1979) Genetic heterogeneity in osteogenesis imperfecta. J Med Genet 16:101–116
7. Van Dijk FS, Sillence DO (2014) Osteogenesis imperfecta: clinical diagnosis, nomenclature and severity assessment. Am J Med Genet A 164A:1470–1481
8. Gautieri A, Uzel S, Vesentini S et al (2009) Molecular and mesoscale mechanisms of osteogenesis imperfecta disease in collagen fibrils. Biophys J 97:857–865
9. Jacobsen CM, Schwartz MA, Roberts HJ et al (2016) Enhanced Wnt signaling improves bone mass and strength, but not brittleness, in the Col1a1(+/mov13) mouse model of type I Osteogenesis Imperfecta. Bone 90:127–132
10. LoMauro A, Fraschini P, Pochintesta S et al (2018) Ribcage deformity and the altered breathing pattern in children with osteogenesis imperfecta. Pediatr Pulmonol 53:964–972

11. Basel D, Steiner RD (2009) Osteogenesis imperfecta: recent findings shed new light on this once well-understood condition. Genet Med 11:375–385
12. Cohn DH, Byers PH (1990) Clinical screening for collagen defects in connective tissue diseases. Clin Perinatol 17:793–809
13. Dalgleish R (1997) The human type I collagen mutation database. Nucleic Acids Res 25:181–187
14. Dalgleish R (1998) The human collagen mutation database. Nucleic Acids Res 26:253–255
15. Beck K, Chan VC, Shenoy N et al (2000) Destabilization of osteogenesis imperfecta collagen-like model peptides correlates with the identity of the residue replacing glycine. Proc Natl Acad Sci U S A 97:4273–4278
16. Rauch F, Lalic L, Roughley P et al (2010) Genotype-phenotype correlations in nonlethal osteogenesis imperfecta caused by mutations in the helical domain of collagen type I. Eur J Hum Genet 18:642–647
17. Casanova EL, Gerstner Z, Sharp JL et al (2018) Widespread genotype-phenotype correlations in intellectual disability. Front Psychiatry 9:535–546
18. Cerkauskaite A, Cerkauskiene R, Miglinas M et al (2019) Genotype(-)Phenotype correlation in a new fabry-disease-causing mutation. Medicina (Kaunas) 55:122
19. Ries M, Gal A, Mehta A et al (2006) Genotype-phenotype correlation in Fabry disease. Oxford Pharma Genesis, Chapter 34
20. Bayat A (2002) Science, medicine, and the future: bioinformatics. BMJ 324:1018–1022
21. Pereira R, Oliveira J, Sousa M (2020) Bioinformatics and computational tools for next-generation sequencing analysis in clinical genetics. J Clin Med 9:132–162
22. Ardeshir B (2002) Science, medicine, and the future: bioinformatics. BMJ 324:1018–1023
23. Tanjo T, Kawai Y, Tokunaga K (2021) Practical guide for managing large-scale human genome data in research. J Hum Genet 66:39–52
24. Doolittle RF (2010) The roots of bioinformatics in protein evolution. PLoS Comput Biol 6:e1000875
25. Hagen JB (2000) The origins of bioinformatics. Nat Rev Genet 1:231–236
26. Ouzounis CA, Valencia A (2003) Early bioinformatics: the birth of a discipline—a personal view. Bioinformatics 19:2176–2190
27. Tenny S, Hoffman MR (2023) Odds ratio. StatPearls, Treasure Island (FL)
28. Szumilas M (2010) Explaining odds ratios. J Can Acad Child Adolesc Psychiatry 19:227–229
29. Andrade C (2015) Understanding relative risk, odds ratio, and related terms: as simple as it can get. J Clin Psychiatry 76:e857-861
30. Cummings, (2009) The relative merits of risk ratios and odds ratios. Arch Pediatr Adolesc Med 163:438–445
31. Blockeel H, Devos L, Frenay B et al (2023) Decision trees: from efficient prediction to responsible AI. Front Artif Intell 6:1124553
32. Becker T, Rousseau AJ, Geubbelmans M et al (2023) Decision trees and random forests. Am J Orthod Dentofacial Orthop 164:894–897
33. Podgorelec V, Kokol P, Stiglic B, etal, (2002) Decision trees: an overview and their use in medicine. J Med Syst 26:445–463
34. Benson DA, Cavanaugh M, Clark K et al (2013) GenBank. Nucleic Acids Res 41:D36-42
35. den Dunnen JT, Antonarakis SE (2001) Nomenclature for the description of human sequence variations. Hum Genet 109:121–124
36. Sun X, Chai Y, Wang Q et al (2015) A natural interruption displays higher global stability and local conformational flexibility than a similar gly mutation sequence in collagen mimetic peptides. Biochemistry 54:6106–6113
37. Bodian DL, Madhan B, Brodsky B et al (2008) Predicting the clinical lethality of osteogenesis imperfecta from collagen glycine mutations. Biochemistry 47:5424–5432
38. Myllyharju J, Kivirikko KI (2001) Collagens and collagen-related diseases. Ann Med 33:7–21

39. Thiagarajan G, Li Y, Mohs A et al (2008) Common interruptions in the repeating tripeptide sequence of non-fibrillar collagens: sequence analysis and structural studies on triple-helix peptide models. J Mol Biol 376:736–748
40. Sun X, Liu S, Yu W et al (2016) CD and NMR investigation of collagen peptides mimicking a pathological Gly-Ser mutation and a natural interruption in a similar highly charged sequence context. Protein Sci 25:383–392

Chapter 4
Synthesis of Collagen Mimetic Peptides

Abstract Collagen presents formidable challenges for direct examination using biophysical techniques due to its considerable size, limited solubility, and intricate hierarchical arrangement. To surmount these obstacles, collagen mimetic peptides, comprising repetitive (Gly-X-Y)$_n$ sequences, have emerged as invaluable tools for probing the sequential and structural fundamentals of collagen's functionality and pathogenesis. This chapter is dedicated to elucidating the synthesis of collagen mimetic peptides, commencing with the generation of triple-helical peptide models of collagen and an exposition on solid-phase peptide synthesis. It covers various aspects including solid supports, terminal protective groups, side chain protection, activating agents, and chromogenic reagents. General procedures for solid-phase peptide synthesis are delineated alongside specific methodologies for crafting collagen mimetic peptides such as (Gly-Pro-Hyp)$_{10}$, FAM-modified (Gly-Pro-Hyp)$_7$Gly, and DOTA-modified (Gly-Hyp-Hyp)$_{10}$. Technical hurdles encountered in solid-phase peptide synthesis, such as incomplete synthesis and diminished efficiency, are addressed. Moreover, innovative synthetic techniques like native chemical ligation, ynamides-mediated peptide synthesis, and enzyme-catalyzed peptide synthesis are introduced.

Keywords Collagen mimetic peptides · Solid-phase peptide synthesis · Synthetic protocols · Triple helical peptide models

4.1 Triple-Helical Peptide Models of Collagen

Collagen, with its substantial size, low solubility, repetitive sequence, and intricate hierarchical structure, poses significant challenges for direct investigation using biophysical techniques [1–6]. In order to overcome these inherent difficulties, scientists have developed collagen mimetic peptides capable of forming the unique triple helix structure, allowing for the characterization of collagen's molecular structure and biological function [7, 8]. These peptides adhere to the strict (Gly-X-Y)$_n$ sequence requirement and possess a high content of imino acids, enabling the formation of

© Springer Nature Singapore Pte Ltd. 2024

J. Xiao, *Collagen Mimetic Peptides and Their Biophysical Characterization*,
Springer Series in Biomaterials Science and Engineering 20,
https://doi.org/10.1007/978-981-10-1941-8_4

Fig. 4.1 Triple helical peptide model of MMPs cleavage sites in type III collagen

stable triple helices. Incorporating the Gly-Pro-Hyp tripeptide sequences, known for their high stability, at one or both ends has become a widely adopted strategy to promote homotrimer formation. This approach has been successfully utilized in constructing a peptide model of the matrix metalloproteinases (MMPs) cleavage site in type III collagen (Fig. 4.1) [9]. Additionally, alternative strategies, including covalent linkage to enforce the selection of three chains and the utilization of electrostatic interactions within the $(\text{Gly-X-Y})_n$ sequences to guide desired self-assembly, have been developed for heterotrimer peptides [10–12].

Triple-helical peptides have emerged as a valuable tool for investigating the structural effects of sequence variations in collagen [13–17]. Studies employing host–guest peptides have revealed that, Arg, commonly found in the Y position of natural collagen, forms the most stable triple helix following Hyp, while negatively charged amino acids in the X position display a strong propensity for stable triple helix formation, akin to Pro. Investigations into triple-helical peptides have demonstrated that the substitution of Gly with different residues leads to significant peptide destabilization, with the stability loss ranking as Ala $\leq$ Ser $<$ Cys $<$ Arg $<$ Val $<$ Glu $\leq$ Asp. X-ray and NMR studies of peptide models carrying Gly-Ala and Gly-Ser mutations have indicated localized perturbations in dihedral angles and hydrogen bonding at the mutation sites. Moreover, introducing Gly mutations into $(\text{POG})_{10}$ peptides has resulted in considerable delays in the folding time. Furthermore, peptide models incorporating natural interruptions have exhibited localized conformational changes [18]. These peptide studies have made substantial contributions to our understanding of how sequence variations influence the stability, conformation, and folding of the triple helix.

Triple-helical peptides offer a sequence-controllable system to investigate the function of collagen. Farndale and colleagues have developed a comprehensive toolkit of over 100 triple helical peptides, covering the entire triple helix domain of type II and type III collagen. They revealed multiple integrin-binding sites in collagen, characterized by a common GXX'GER motif, with different GXX'GER sequences exhibiting distinct affinities for integrins [19]. Fields and colleagues have systematically compared the capacity of various MMPs to degrade different triple helical peptides and discovered that specific amino acid types at certain sites could regulate the extent of collagen degradation [20]. These triple-helical peptide models provide valuable opportunities to explore the underlying sequential and structural basis of collagen function and pathology.

4.2 Introduction to Solid Phase Peptide Synthesis

Peptide synthesis has made remarkable strides since the pioneering synthesis of the first dipeptide in 1903 [21, 22]. Initially, coupling two amino acids was achievable via acyl chlorides, but longer peptides were challenging due to the lack of suitable amino-protecting groups. The introduction of the benzyloxycarbonyl-protecting group in 1932 by Bergmann and Zervas revolutionized peptide synthesis. However, limited sequencing techniques at the time delayed the historical synthesis of peptide oxytocin until 1953, recognized with the Nobel Prize in 1955. Since then, numerous peptides of significant biological importance and economic value, such as insulin and glucagon, have been successfully synthesized using similar approaches.

However, the chemical synthesis of longer peptides in solution remains very laborious and challenging, since the method requires many repetitive steps of coupling, deprotection, isolation and purification. The presence of both amino and carboxyl groups in each amino acid can lead to unwanted side reactions, resulting in the formation of considerable byproducts. Moreover, when multiple amino acids are involved in the reaction, peptides with various amino acid sequences can be generated, making it difficult to obtain the desired peptide with the intended sequence.

To overcome these challenges, the solid-phase peptide synthesis (SPPS) technique was developed by Merrifield, for which he was awarded the Nobel Prize in 1984. SPPS offers several critical advantages over solution-phase synthesis. Firstly, coupling reactions can be performed rapidly and with high yield using an excess of activated amino acid derivatives. Secondly, the purification process is greatly simplified as unreacted amino acids and reagents can be easily removed through washing steps, resulting in highly pure intermediates [22].

SPPS has established itself as the gold standard method for peptide and protein synthesis, offering versatility in the synthesis of challenging peptides that are hard to express in bacteria or contain modifications such as unnatural amino acids, D-amino acids, or alterations in the backbone or side chains. The continuous refinement of SPPS techniques, including the development of improved solid supports, mild reaction conditions, and efficient coupling reagents, has contributed to its remarkable success. With the availability of modified commercial resins and enhanced N-terminal amino group protection methods, the synthesis of medium-sized peptides can be achieved conveniently through manual or automated SPPS. This section provides a comprehensive overview of the key principles in SPPS.

4.2.1 Solid Supports

SPPS relies on the utilization of appropriate solid supports, which serve as a platform for peptide synthesis reactions. These solid supports, typically inert and insoluble polymers, possess the ability to selectively protect and deprotect active groups [23, 24]. They must exhibit extensive swelling properties to facilitate rapid reagent penetration and enable the attachment of the initial amino acid. The effectiveness of the solid supports depends on factors such as the composition of the materials, the degree of cross-linking, and the specific linker and handle employed.

Significant advancements have been made in the development of solid supports, leading to the availability of various commercial resins [25]. These resins can be categorized into three types based on the mode of linking the first amino acid to the resin. Wang resin and 2-chlorotrityl chloride resin allow the amino acids to be linked to the resin in the form of ester bonds; MBHA Resin and Knorr Resin allow the amino acids to be linked to the resin in the form of amide bonds; 4-Methoxytrityl chloride resin and 4-Methyltrityl chloride resin allow the amino acids to be linked to the resin in the form of alcohol bonds. Furthermore, resins can be classified as either polar or nonpolar, based on their swelling properties in different solvents.

Wang resin, widely recognized as the preferred resin for acid substrates, is a polystyrene resin featuring a 4-hydroxybenzyl alcohol linker [25]. The linker is connected to the resin via a phenyl ether bond, while the substrate is attached to the linker through a benzylic ester or ether bond. This linkage exhibits excellent stability under diverse reaction conditions but can be readily cleaved by mild treatment with trifluoroacetic acid (TFA). In contrast, 2-chlorotrityl chloride resin addresses the issue of racemization during the attachment of the initial amino acid, and its substantial steric hindrance also impedes diketopiperazine formation.

4.2.2 N-Terminal Protecting Groups

Amino acids contain reactive functional groups at their N-terminal, C-terminal, and side chains. In peptide synthesis, excess protected amino acids are used in each step to ensure high yields. Without N-terminal protection of the added amino acid, undesired self-coupling reactions can occur, impeding the successful synthesis of the desired peptide [26]. Therefore, an additional deprotection step is required to remove the N-terminal protecting group before each subsequent coupling step in the synthesis cycle.

Protection groups can be classified into two categories: temporary protection of the α-amino group and permanent protection to block side-chain functionalities of amino acids [21]. Permanent protecting groups are removed only during the final cleavage of the synthesized peptide from the resin, thus they need to withstand the repeated cleavage steps during synthesis. Premature removal of permanent protecting groups can lead to the formation of undesired by-products. The tert-butyloxycarbonyl (t-Boc) group and fluorenylmethyloxycarbonyl (Fmoc) group are the two most commonly used temporary protecting groups in SPPS.

In the t-Boc protection scheme, Boc protecting groups provide temporary protection of the α-amino group of the amino acids, which can be easily removed under moderate conditions (50% TFA in DCM) [27]. On the other hand, benzyl-based protecting groups offer more prolonged protection of side chains but necessitate the use of strong acids such as HF or TFMSA for their removal. It is essential to acknowledge that both Boc and benzyl-based protecting groups are sensitive to acid, and therefore, they do not provide a fully orthogonal protection scheme. The addition of t-Boc groups to amino acids can be achieved using t-Boc anhydride and an appropriate base. This approach proves particularly advantageous in the synthesis of non-natural peptide analogs that are sensitive to basic conditions.

In the Fmoc scheme, the α-amino group of the amino acids is protected by the base-labile Fmoc group, while the side chains are protected using acid-labile protecting groups such as tert-butyl (t-Bu) or trityl (Trt) groups. The Fmoc method employs piperidine (20–50%) in DMF as a base for the deprotection of the Fmoc group [28]. The high stability of Fmoc under acidic conditions and its mild deprotection under slightly basic conditions allow the use of acid-labile protecting groups, such as Boc and benzyl groups, to protect the side chains of amino acids. This results in an orthogonal protection system, where the side chain protecting groups can be selectively removed without affecting the N-terminal protection, and vice versa. This approach is particularly advantageous when specific modifications or functionalizations are required on the side chains of amino acid residues.

4.2.3 C-Terminal Protecting Groups

SPPS generally does not necessitate the protection of C-terminal carboxyl groups, as they are typically linked to the solid support through a linker, which serves as a protective group. However, in certain cases where the peptide is attached to the resin via an amino acid side chain or when the synthesis proceeds in the reverse N–C direction, C-terminal protection becomes necessary [29, 30]. An example of this is the synthesis of glutathione, which involves the protection of the α-carboxyl group of glutamic acid.

The protection of the α-carboxylic acid group can be achieved through various reactions: (1) tert-butyl protection by treating an α-amino-free or protected amino acid with isobutene under acidic conditions; (2) reaction of an α-amino-free amino acid with an alcohol in acidic conditions; (3) reaction of an α-amino-protected amino acid with the corresponding halide in the presence of base or as a cesium salt.

4.2.4 Protection of Side Chains

Side-chain protection is necessary for certain amino acids that possess reactive functional groups during SPPS. Amino acids such as Asp, Glu, Cys, Arg, Lys, His, Ser, and Thr typically require side-chain protection, while other amino acids such as Asn, Gln, Trp, and Met may require protection depending on the specific peptide sequence, coupling and cleavage strategies, and desired modifications. Side-chain protection groups are considered semi-permanent, as they are removed during resin cleavage to yield the final unprotected peptide. Peptide synthesis utilizes a variety of side-chain protection strategies, which are selected based on factors like peptide sequence, length, synthetic scale, and post-synthesis modifications (Table 4.1).

To protect the side-chain carboxylic groups of Asp and Glu during peptide synthesis, commonly used protection groups include t-Bu in the Fmoc/t-Bu strategy and cyclohexyl (cHx) in the Boc/Bn strategy [31–33]. These protection groups are essential to prevent activation and unwanted branching reactions, as well as to minimize the formation of aspartimide, which can lead to the generation of undesired

Table 4.1 Side-chain protecting groups for Fmoc amino acids

Amino acids	Functional group	Protecting group
Asp (D)	Carboxyl	OtBu, OAI
Glu (E)	Carboxyl	OtBu, OAI
Cys (C)	Sulfhydryl	Trt, Acm, t-Bu, StBu
Arg (R)	Guanidino	Pbf, Mtr, Pmc
Lys (K)	Amino	Boc, Aloc, Fmoc
His (H)	Imidazole	Trt, Boc
Ser (S)	Hydroxyl	t-Bu
Thr (T)	Hydroxyl	t-Bu
Asn (N)	Amide	Trt, Mbh, Tmob
Gln (Q)	Amide	Trt, Mbh, Tmob
Trp (W)	Indole	Boc
Orn (O)	Amino	Boc
Tyr (Y)	Phenol	t-Bu
Ala (A)	–	–
Gly (G)	–	–
Ile (I)	–	–
Leu (L)	–	–
Met (M)	–	–
Phe (F)	–	–
Pro (P)	–	–
Val (V)	–	–

β-peptide byproducts. The cHx group, in particular, is known for its effectiveness in preventing aspartimide formation [31].

The side chain of Cys necessitates protection due to its nucleophilic thiol group, which is vulnerable to acylation, alkylation, and oxidation by air. Despite protection, Cys can still undergo various side reactions, and racemization is a concern during the coupling reaction. In the Fmoc/t-Bu strategy, commonly used protection groups for Cys are Acm or Trt, while the Boc/Bn strategy utilizes Bn/Meb or Acm groups. The selection of protection groups influences whether disulfides or free thiols are obtained after peptide synthesis [34].

Protection of the guanidino group in Arg is crucial to prevent deguanidination and the formation of Orn and δ-lactam. The commonly used strategy involves sulfonyl protection of the ω-amino function. In the Fmoc/t-Bu strategy, popular protecting groups for Arg are Pbf (pentamethyl-2,3-dihydrobenzofuran-5-sulfonyl) and Pmc (2,2,5,7,8-pentamethylchroman-6-sulfonyl), while Tos (p-toluenesulfonyl) is commonly used in the Boc/Bn strategy. It is worth noting that these protection groups are highly stable, and their removal can be challenging, especially for peptides containing multiple arginine residues [35].

Protection of the side chains of Lys is essential to prevent acylation and the formation of undesirable branched peptides. The α-amino functionality protection groups can also be employed for side-chain protection, although removal of ω-amino protection is more challenging due to its higher basicity. The Boc group is commonly employed for side-chain protection of Lys in the Fmoc/t-Bu strategy, while the 2-chlorobenzyloxycarbonyl (Cbz) group is frequently used in the Boc/Bn strategy [36].

4.2.5 Activating Groups

Synthetic peptide coupling requires the activation of the C-terminal carboxylic acid on the incoming amino acid to speed up the reaction. Carbodiimides, such as dicyclohexylcarbodiimide (DCC) or diisopropylcarbodiimide (DIC), are commonly used activating reagents that react with the carboxyl group to form a highly reactive O-acylisourea intermediate. This intermediate is readily replaced by nucleophilic attack from the amino group on the N-terminus of the growing peptide chain, resulting in the formation of the nascent peptide bond. DCC acts as a catalyst by enhancing the electrophilicity of the carboxylate group, thereby activating the negatively charged oxygen and facilitating its departure as a better leaving group during peptide coupling. However, the high reactivity of carbodiimides, such as DCC, can lead to undesirable racemization of the amino acid during peptide synthesis.

To address the issue of racemization, alternative coupling reagents have been developed, such as 1-hydroxybenzotriazole (HOBt) and 1-hydroxy-7-azabenzotriazole (HOAt). These reagents react with O-acylurea intermediates to form active esters that are less reactive, minimizing the risk of racemization. Additionally, other coupling agents like benzotriazol-1-yloxytris(dimethylamino)phosphonium hexafluorophosphate (BOP) and 2-(1H-benzotriazol-1-yl)-1,1,3,3-tetramethyluronium hexafluorophosphate (HBTU) have been developed to mitigate side reactions associated with carbodiimides. These coupling agents require activating bases to facilitate amino acid coupling.

4.2.6 Chromogenic Reagents

The efficiency of coupling reaction is critical for the synthesis of peptides with desired lengths and appropriate purity. Chromogenic reagents that can monitor the degree of condensation of amino acids are thus necessitated in order to ensure the completion of each coupling step. Ninhydrin and tetrachlorobenzene/acetaldehyde are the two most commonly used chromogenic reagents in peptide synthesis.

(1) Ninhydrin reaction

The ninhydrin reaction is a highly sensitive method for detecting free amino groups on the resin. It can detect as low as 5 μmol/g, allowing identification of 1% unreacted amino groups and ensuring a condensation level of 99%. The color observed in the ninhydrin reaction depends on the specific terminal amino acid with the free amino group. For instance, aspartic acid exhibits a weak blue color, while asparagine appears as a light brown color.

(2) Tetrachlorobenzene/acetaldehyde chromogenic reaction

The tetrachlorobenzene/acetaldehyde chromogenic method is a commonly used technique for detecting primary and secondary amines on the resin. Two original solutions, prepared with DMF as the solvent, consist of a tetrachlorobenzene solution and an acetaldehyde solution, both typically with a concentration of 2%. To initiate the chromogenic reaction, a small amount of resin is added to a mixture of equal volumes of the tetrachlorobenzene and acetaldehyde solutions. After incubating at room temperature for 5 min, any potential color change indicates the presence of unbound amino groups. Resin without unbound amino groups remains colorless or light yellow, while resin containing free primary or secondary amine groups will display a distinct dark blue or green coloration.

4.3 General Protocols for Solid Phase Peptide Synthesis

Solid phase peptide synthesis generally starts from the C-terminus toward the N-terminus of the peptide in order to avoid the risk of amino acid racemization [37]. The basic steps of SPPS are shown as Fig. 4.2.

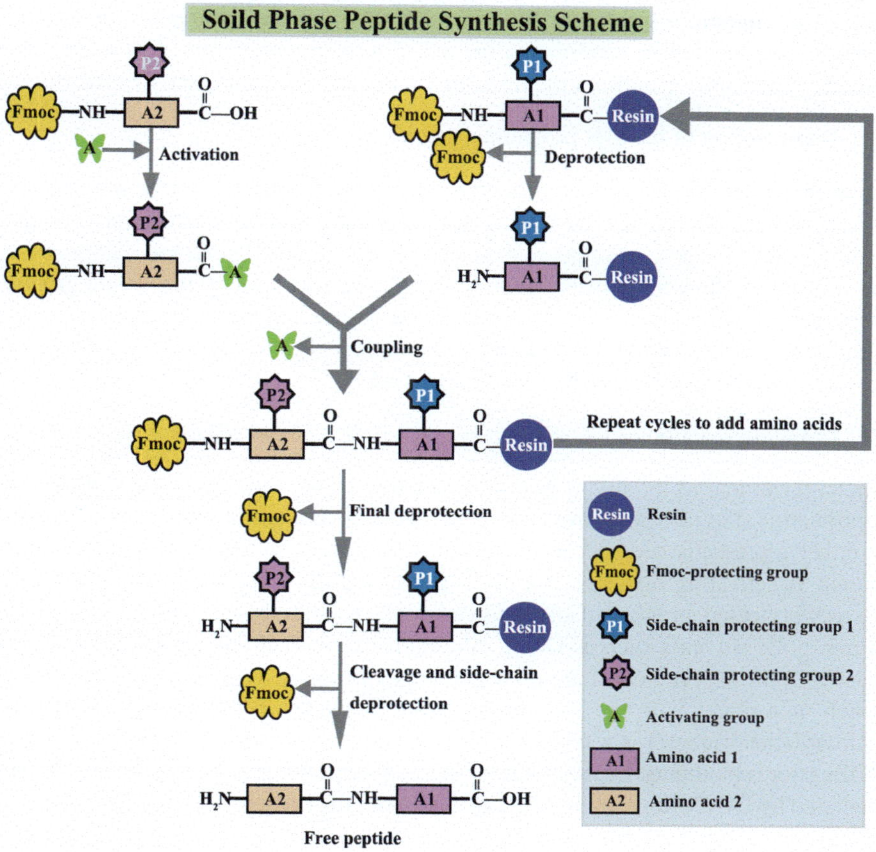

Fig. 4.2 Illustrative scheme of solid phase peptide synthesis

(1) Resin swelling

Achieving optimal resin swelling is crucial to accommodate the growing peptide chain. The degree of crosslinking, usually ranging from 1 to 2%, plays a critical role in determining the swelling properties. Resin swelling is influenced by the polarity of solvents used; polystyrene resins, for example, exhibit significant swelling in nonpolar solvents like dichloromethane (DCM) and dioxane, moderate swelling in N,N-dimethylformamide (DMF), and limited swelling in polar solvents like alcohols and water. The process of resin swelling typically takes approximately 1–3 h.

(2) Loading the first α-N protected amino acid to the resin

The initial α-N protected amino acid is conjugated through its carboxyl group to a hydroxyl (or chloro) or amino resin, resulting in the formation of either an ester- or amide- linked peptide. This process ultimately yields a peptide with a C-terminal acid or a C-terminal amide. The condensation reaction is carried out using a condensation reagent at ambient temperature for several hours, and the progress is monitored using a color test before terminating the reaction upon completion. Any remaining unreacted hydroxyl groups on the resin are then blocked to complete the process.

(3) Deprotection of the α-N protecting group

Upon the formation of each peptide bond, it becomes imperative to eliminate the α-N protecting group utilized for temporary safeguarding of the α-N amino groups. This deprotection procedure necessitates gentle conditions that do not compromise the stability or elongation of the peptide chain. The two primary protecting groups employed are Boc and Fmoc. Boc protecting groups are typically cleaved using trifluoroacetic acid (TFA), while Fmoc deprotection is achieved using a mixture of hexahydropyridine and dimethylformamide (DMF).

(4) Coupling reaction

Upon attachment of the first α-N protected amino acid to the resin and subsequent deprotection, the next α-N protected amino acid is coupled through the utilization of either a pre-activated species or in situ activation. Condensation reagents, also known as activating reagents, can directly catalyze the coupling reaction between the α-N protected amino acid and the free amino group of the growing peptide chain in non-polar solvents such as DCM. Alternatively, condensation reagents can first react with the α-N protected amino acid to generate an activated ester or anhydride, which then reacts with the free amino group to form a peptide bond. HOBt (1-hydroxybenzotriazole) is a widely used activating agent, and the formation of the HOBt ester is facilitated by the reaction between the protected amino acid and HOBt, mediated by DCC (dicyclohexylcarbodiimide) or DIC (diisopropylcarbodiimide). A mixture of DCM and DMF (dimethylformamide) is commonly employed for both the activation and coupling steps.

(5) Repetitive cycles of steps (3) and (4) until the synthesis of the whole sequence

After loading the first amino acid, the desired peptide sequence is synthesized one amino acid after the other in a linear fashion from the C-terminus to the N-terminus by repetitive cycles of α-N deprotection and amino acid coupling reactions.

(6) Peptide cleavage from the resin

In Boc-SPPS, the target peptide is cleaved from the resin using anhydrous HF, which requires a specialized container due to its highly corrosive nature. In Fmoc-SPPS, the target peptide is cleaved using a mild TFA solution, followed by the addition of ether to the filtrate. The target peptide is then precipitated, centrifuged, and purified by HPLC to obtain the target peptide.

4.4 Synthesis of Collagen Mimetic Peptides

SPPS has been utilized to synthesize collagen mimetic peptides (CMPs) since 1960's. The first synthesized peptide contains sequence of $(PPG)_{10}$, which shows similar temperature dependence of optical rotation when compared with natural collagen. However, unlike natural collagen, synthetic CMPs can form single crystals, enabling consistent diffraction patterns with natural collagen and providing valuable structural insights at the molecular level, highlighting the remarkable potential of triple helical peptides as superior mimics for collagen.

Fmoc-based SPPS has become the predominant strategy for synthesizing CMPs in recent times. The sequential coupling of imino acids is one of the main challenges for the synthesis of imino acid-rich triple helical peptides. To address these difficulties, various approaches have been developed, such as extending the amino acid coupling time, using a mixture of diaza(1,3)-bicyclo[5.4.0]undecane and piperidine for Fmoc deprotection, coupling triplets instead of individual amino acids, and implementing double coupling. Manual couplings with real-time monitoring of the completion of each synthesis step together with the use of HATU as a coupling agent has also proved beneficial in synthesizing CMPs with challenging sequences.

4.4.1 *Synthesis of (Gly-Pro-Hyp)$_{10}$*

Peptide $(Gly\text{-}Pro\text{-}Hyp)_{10}$ was synthesized via standard Fmoc-based SPPS using 2-chlorotrityl chloride resin (0.8 mmol/g loading). Stepwise couplings of amino acids were carried out using Fmoc-amino acids (4 eq.), DIEA (6 eq.), HBTU (4 eq.), and HOBt (4 eq.). Resin washes with DMF (3 × 10 mL) and DCM (2 × 10 mL) followed each coupling step, and Fmoc deprotection was achieved with 20% (v/v) piperidine in DMF. The coupling reaction and Fmoc deprotection were monitored using the Chloranil test. After the completion of the synthesis of peptide sequence $(Gly\text{-}Pro\text{-}Hyp)_{10}$, the resin was treated with TFA/H_2O/TIS (90:5:5) for 2.5 h to deprotect the side-chain protecting group and cleave the peptide from the resin. The peptide was harvested by precipitation with cold Et_2O. Crude products were collected by re-suspending in cold Et_2O, followed by sonication and centrifugation. Subsequently, purification was performed using reverse-phase HPLC on a C18 column with a water-acetonitrile gradient. The identity of the peptide was confirmed using MALDI-TOF MS with a-cyano-4-hydroxycinnamic acid as the matrix. The peptide was lyophilized and stored at –20 °C.

4.4.2 Synthesis of FAM Modified (Gly-Pro-Hyp)$_7$Gly

FAM modified (Gly-Pro-Hyp)$_7$-Gly peptide was synthesized by standard Fmoc-based SPPS using 2-chlorotrityl chloride resin (0.8 mmol/g loading). Stepwise couplings of amino acids were performed using Fmoc-amino acids (4 eq.), DIEA (6 eq.), HBTU (4 eq.) and HOBt (4 eq.). Resin was washed by DMF (3 × 10 mL) and DCM (2 × 10 mL) after each step of coupling, and the Fmoc protecting group was removed with 20% (v/v) piperidine in DMF. Chloranil test was used to check the status of coupling reaction and Fmoc deprotetction. After the completion of the synthesis of peptide sequence (Gly-Pro-Hyp)$_7$-Gly, the mixture of FAM (5 eq.), HBTU (5 eq.) and DIEA (10 eq.) in DMF was added to the resin, and incubated for 24 h to conjugate FAM to the N-terminal of the peptide. The resin was treated with TFA/H_2O/TIS (90:5:5) for 2.5 h to deprotect the side-chain protecting group and cleave the peptide from the resin. The peptide was harvested through precipitation with cold Et_2O. Afterward, the crude product was collected by re-suspending it in cold Et_2O, followed by sonication and centrifugation. The subsequent purification process involved reverse-phase HPLC on a C18 column using a water-acetonitrile gradient. Following the confirmation of the peptide's identity using MALDI-TOF MS, it was lyophilized and stored at –20 °C.

4.4.3 Synthesis of DOTA Modified (Gly-hyp-Hyp)$_{10}$

The DOTA-modified (Gly-Hyp-Hyp)$_{10}$ peptide was synthesized using the standard Fmoc SPPS method with rink amide resin (0.345 mmol/g loading). The amino acids were coupled stepwise using Fmoc-amino acids (4 eq.), DIEA (6 eq.), HBTU (4 eq.), and HOBt (4 eq.). After each coupling step, the resin was washed with DMF and DCM, and the Fmoc protecting group was removed using 20% (v/v) piperidine in DMF. The chloranil test was employed to check the status of the coupling reaction and Fmoc deprotection. Upon completing the peptide sequence synthesis, a mixture of DOTA (tBu)$_3$ (10 eq.), HOBt (10 eq.), HBTU (10 eq.), and DIEA (16 eq.) in DMF was added to the resin and incubated at 37 °C for 24 h to conjugate DOTA with the N-terminal of the peptide. For the final deprotection and cleavage of the peptide from the resin, TFA/H_2O/TIS (95:2.5:2.5) was applied for 2.5 h. Crude products were obtained through re-suspension in cold Et_2O, sonication, and centrifugation, and further purified using reverse-phase HPLC on a C18 column. The peptide's identity was confirmed through MALDI-TOF MS, and it was subsequently lyophilized and stored at –20 °C.

4.5 Technical Challenges for Solid Phase Peptide Synthesis

Despite continuous improvements in SPPS, practical peptide synthesis encounters various challenges:

(1) Incomplete peptide synthesis

Incomplete peptide synthesis may arise due to intermolecular aggregation of hydrophobic amino acids on the resin, leading to byproducts closely resembling the target peptide but lacking a few amino acids. This similarity makes their separation and purification challenging. Higher resin substitution often results in increased aggregation, so reducing resin substitution through methods like using polar solvents or heating the reaction solution can mitigate this issue.

(2) Limited peptide length

SPPS is generally limited to synthesizing medium-sized peptides due to slow coupling reactions and low peptide concentrations, making the synthesis of longer peptides challenging.

(3) Low yield

The deprotected second amino acid may bind the ester bond of the resin, resulting in a low yield of the target peptide, especially when the first amino acid is proline or N-methylglycine.

(4) Decreased synthesis efficiency

Peptides rich in disulfide bonds and basic amino acids can be difficult to synthesize, and as the number of amino acids in the peptide increases, the efficiency of SPPS may decrease.

(5) Lengthy synthesis time

The addition of each amino acid in SPPS can take 20–120 min, and difficult peptides may require even longer synthesis times.

(6) Toxic reagents

Some chemical reagents used in SPPS are toxic, which can raise safety concerns and necessitate the development of more environmentally friendly and safer reagents.

4.6 Novel Synthetic Methods of Peptides

In the dynamic field of proteomics, the fascinating discovery of proteins with novel structures and functions has ignited a profound interest in synthesizing peptides with intricate and extended sequences. This shift in focus has created a surge in demand for synthesizing peptides by coupling multiple peptide segments, which is a departure from traditional peptide synthesis mainly involving simple amino acid coupling. To address these challenges and cater to the evolving needs of innovative peptides, various cutting-edge synthetic methods have been developed, including native chemical ligation, ynamides-mediated peptide synthesis, and enzyme-catalyzed peptide synthesis. These advancements mark an exciting era in the quest for exploring the vast potential of peptides in proteomics research and beyond.

4.6.1 Native Chemical Ligation for Peptide Synthesis

Native chemical ligation (NCL) is widely acknowledged as one of the most efficient methods for synthesizing long peptides by joining two or more peptide fragments in an aqueous solution. Notably, NCL allows the use of unprotected peptide fragments, eliminating the need for extra enzymes or chemically coupled reagents. In 1994, Kent and co-workers introduced the first NCL method, utilizing disulfide bonds between cysteines. However, the practicality of this Cys-based NCL approach has been limited by the rarity of cysteine residues in native proteins. In response to this constraint, novel thioester-mediated and imine-mediated NCL strategies have recently emerged, providing alternative pathways for synthesizing long peptides without relying on cysteine residues.

Low and colleagues have achieved a significant milestone by completely chemically synthesizing cytochrome b562, which consists of a total of 106 amino acids [38]. This impressive synthesis was accomplished through thioester-mediated coupling of two unprotected peptide fragments, comprising 63 and 42 amino acids, respectively. To enable amide bond formation during ligation, they strategically introduced a removable auxiliary group (1-phenyl-2-mercaptoethyl) at the N-terminal amino group of the 42 amino acid peptide segment. Subsequent treatment with anhydrous hydrogen fluoride led to the formation of a new amide bond while simultaneously removing the auxiliary group, resulting in a full-length unmodified peptide product, with a Gly residue acting as the binding site.

TAM and their collaborators successfully synthesized a series of analogues of bactenecin 7, an antimicrobial peptide comprising 59 amino acid residues, through imine-mediated linkage of two segments (Fig. 4.3) [39]. Initially, four peptide segments were synthesized, including a 24-amino acid-containing acyl segment 18 with a glycoaldehyde ester at the C-terminus, and three 35-amino acid-containing segments featuring Ntn-Ser (19a), Ntn-Thr (19b), and Ntn-Cys (19c) at the N-terminus, respectively. Subsequently, the peptide segments 18, 19a, 19b, and 19c were coupled together via the creation of a monosubstituted pseudoproline bond, leading to the synthesis of bactenecin 7 analogues 20a, 20b, and 20c, respectively. These advancements expand the repertoire of methods available for creating complex peptides.

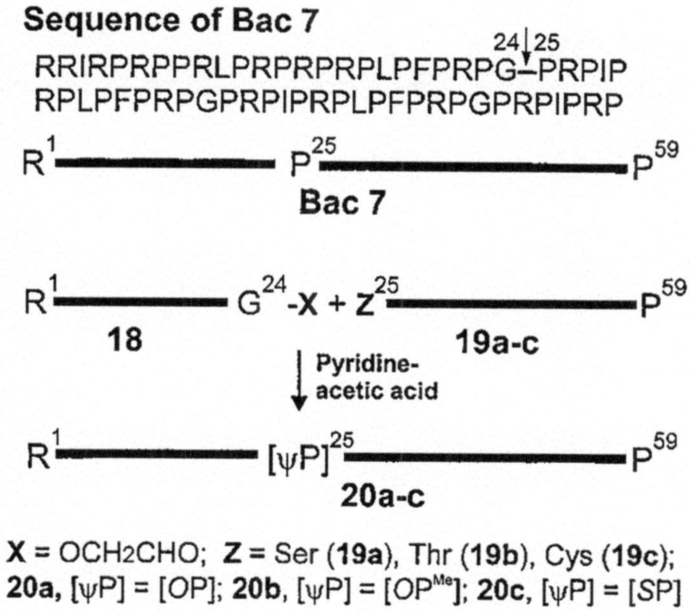

Fig. 4.3 Imine-mediated natural chemical linkage strategy [39]. Reprinted from ref [39], Copyright © 1999, with permission from American Chemical Society

4.6.2 Ynamides-Mediated Peptide Synthesis

Ynamides with an electron-withdrawing group on the nitrogen atom have recently emerged as a promising solution for facilitating the linkage of peptide fragments and effectively addressing the issue of racemization/epimerization during peptide bond formation. A variety of ynamides such as N-methylynemethylsulfonamide (MYMsA) and N-methylynetoluenesulfonamide (MYTsA) have been explored, with MYMsA standing out as particularly accessible and stable. Through a MYMsA-mediated one-pot strategy, Zhao and co-workers successfully synthesized the protected Leu-enkephalin (Fig. 4.4) [40]. This process involved the synthesis of two peptide segments, Fmoc-Tyr(Bzl)-Gly-OH and Fmoc-Gly-L-Phe-L-Leu-OtBu, with no racemization observed. The removal of the Fmoc group from Gly was achieved through incubation at room temperature for 1 h in Et_2NH and MeOH. Subsequently, the dipeptide and tripeptide were linked via MYMsA-mediated peptide bond formation.

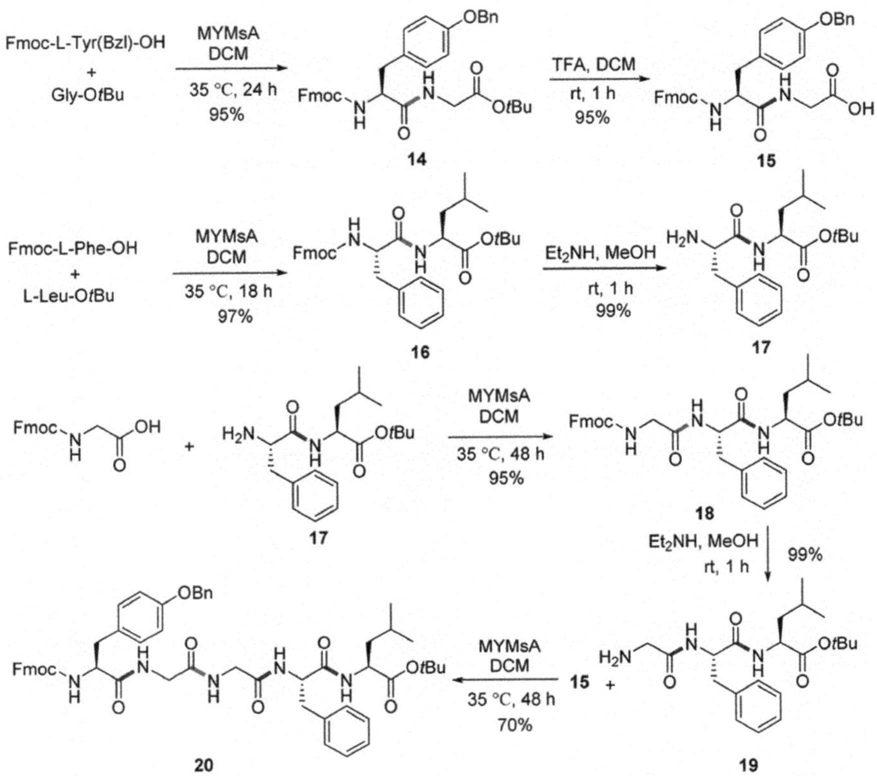

Fig. 4.4 MYMsA as the coupling reagent for the synthesis of protected Leu-Enkephalin [40]. Reprinted from ref [40], Copyright © 2016, with permission from American Chemical Society

4.6.3 Enzyme-Catalyzed Peptide Synthesis

Enzymatic synthesis has garnered increasing attention as a versatile and efficient method for peptide synthesis, offering significant advantages such as mild reaction conditions, high selectivity, and the ability to incorporate non-natural amino acids or non-amino acid structures into the peptide backbone, diversifying the peptide repertoire. Among the notable enzymes, Bunga telang plus ligase (Butelase-1), also known as asparaginyl endoprotease, stands out for its remarkable potential in enabling the synthesis of macrocycle peptides (Fig. 4.5) [41]. Butelase-1 has been found to exhibit a specific recognition for the C-terminal tripeptide sequence DHV or NHV, leading to the cleavage of the peptide bond from the N-terminal of H and releasing the HV dipeptide sequence, forming the acylase substrate complex. Subsequently, the linear peptide's N-terminal amino acid attacks the complex, resulting in the formation of the peptide bond with Asp or Asn at the C-terminal, ultimately leading to the production of cyclic peptides.

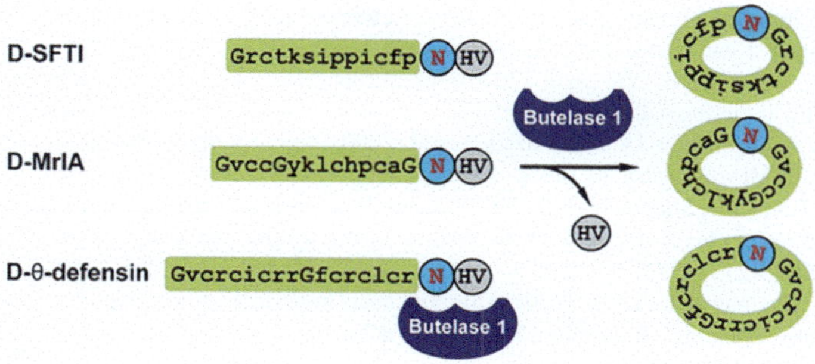

Fig. 4.5 Butelase-mediated macrocyclization of D-amino-acid-containing peptides [41]. Reprinted from ref [41], Copyright © 2016, with permission from John Wiley and Sons

A noteworthy feature is that Butelase-1 can accommodate any natural amino acid except Pro at the first residue of the peptide substrate's N-terminal. However, its exceptional selectivity lies in the strict preference for aliphatic amino acids, particularly favoring Ile, Leu, and Val, in the second position. With these exceptional enzymatic capabilities, Butelase-1 serves as a highly valuable tool for the synthesis of diverse and complex macrocyclic peptides, opening up exciting possibilities for research and therapeutic applications. The continuous advancements in enzymatic synthesis techniques hold great promise for future developments in peptide chemistry and biomedical research.

References

1. Orgel JPRO, Irving TC, Miller A et al (2006) Microfibrillar structure of type I collagen in situ. P Natl Acad Sci USA 103:9001–9005
2. Shoulders MD, Raines RT (2009) Collagen structure and stability. Annu Rev Biochem 78:929–958
3. Ricard-Blum S (2011) The collagen family. Cold Spring Harb Perspect Biol 3:a004978
4. Sorushanova A, Delgado LM, Wu ZN et al (2019) The collagen suprafamily: from biosynthesis to advanced biomaterial development. Adv Mater 31:e1801651
5. Peter F (2008) Collagen: structure and mechanic. Springer, Germany
6. Kielty CM, Grant ME (2002) The collagen family: structure, assembly, and organization in the extracellular matrix. In: Connective tissue and its heritable disorders, pp 159–221
7. Okuyama K, Wu GH, Jiravanichanun N et al (2006) Helical twists of collagen model peptides. Biopolymers 84:421–432
8. Emsley J, Knight CG, Farndale RW et al (2000) Structural basis of collagen recognition by integrin alpha 2 beta 1. Cell 101:47–56
9. Xiao JX, Addabbo RM, Lauer JL et al (2010) Local conformation and dynamics of isoleucine in the collagenase cleavage site provide a recognition signal for matrix metalloproteinases. J Biol Chem 285:34181–34190
10. O'Leary LE, Fallas JA, Hartgerink JD (2011) Positive and negative design leads to compositional control in AAB collagen heterotrimers. J Am Chem Soc 133:5432–5443
11. Renner C, Sacca B, Moroder L (2004) Synthetic heterotrimeric collagen peptides as mimics of cell adhesion sites of the basement membrane. Biopolymers 76:34–47

12. Russell LE, Fallas JA, Hartgerink JD (2010) Selective assembly of a high stability AAB collagen heterotrimer. J Am Chem Soc 132:3242-+
13. Cai XD, Wei WY, Bi YH et al (2020) An intrinsically nontrimerizing peptide probe for specifically targeting pathological collagen in connective tissues. Adv Funct Mater 30:2004532
14. Persikov AV, Ramshaw JA, Kirkpatrick A et al (2000) Amino acid propensities for the collagen triple-helix. Biochemistry 39:14960–14967
15. Beck K, Chan VC, Shenoy N et al (2000) Destabilization of osteogenesis imperfecta collagen-like model peptides correlate with the identity of the residue replacing glycine. P Natl Acad Sci USA 97:4273–4278
16. Xiao JX, Madhan B, Li YJ et al (2011) Osteogenesis imperfecta model peptides: incorporation of residues replacing gly within a triple helix achieved by renucleation and local flexibility. Biophys J 101:449–458
17. Xu T, Zhou CZ, Xiao J et al (2018) Unique conformation in a natural interruption sequence of type XIX collagen revealed by its high-resolution crystal structure. Biochemistry 57(7):1087–1095
18. Sun X, Chai Y, Wang Q et al (2015) A natural interruption displays higher global stability and local conformational flexibility than a similar gly mutation sequence in collagen mimetic peptides. Biochemistry 54:6106–6113
19. Siljander PRM, Hamaia S, Peachey AR et al (2004) Integrin activation state determines selectivity for novel recognition sites in fibrillar collagens. J Biol Chem 279:47763–47772
20. Fields GB (2010) Synthesis and biological applications of collagen-model triple-helical peptides. Org Biomol Chem 8:1237–1258
21. Isidro-Llobet A, Alvarez M, Albericio F (2009) Amino acid-protecting groups. Chem Rev 109:2455–2504
22. Coin I, Beyermann M, Bienert M (2007) Solid-phase peptide synthesis: from standard procedures to the synthesis of difficult sequences. Nat Protoc 2:3247–3256
23. Bayer E (1991) Towards the Chemical Synthesis of Proteins. Angew Chem Int Edit 30:113–129
24. Zalipsky S, Chang JL, Albericio F et al (1994) Preparation and applications of polyethylene glycol-polystyrene graft resin supports for solid-phase peptide-synthesis. React Polym 22:243–258
25. Mcmurray JS (1991) Solid-phase synthesis of a cyclic peptide using fmoc chemistry. Tetrahedron Lett 32:7679–7682
26. Cruz LJ, Beteta NG, Ewenson A et al (2004) One-pot" preparation of N-carbamate protected amino acids via the azide. Org Process Res Dev 8:920–924
27. de la Torre BG, Marcos MA, Eritja R et al (2001) Solid-phase peptide synthesis using N-alpha-trityl-amino acids. Lett Pept Sci 8:331–338
28. Liebe B, Kunz H (1997) Solid-phase synthesis of a tumor-associated sialyl-T-N antigen glycopeptide with a partial sequence of the "tandem repeat" of the MUC-1 mucin. Angew Chem Int Edit 36:618–621
29. Jensen KJ, Alsina J, Songster MF et al (1998) Backbone Amide Linker (BAL) strategy for solid-phase synthesis of C-terminal-modified and cyclic peptides. J Am Chem Soc 120:5441–5452
30. Visintin C, Aliev AE, Riddall D et al (2005) Membrane receptor probes: solid-phase synthesis of biotin-Asp-PEG-arvanil derivatives. Org Lett 7:1699–1702
31. Mergler M, Dick F, Sax B et al (2003) The aspartimide problem in Fmoc-based SPPS. Part I. J Peptide Sci 9:36–46
32. Mergler M, Dick F, Sax B et al (2003) The aspartimide problem in Fmoc-based SPPS. Part II. J Peptide Sci 9:518–526
33. Mergler M, Dick F (2005) The aspartimide problem in Fmoc-based SPPS. Part III. J Peptide Sci 11:650–657
34. West CW, Estiarte MA, Rich DH (2001) New methods for side-chain protection of cysteine. Org Lett 3:1205–1208
35. Ramage R, Green J, Blake AJ (1991) An acid labile arginine derivative for peptide-synthesis—N(G)-2,2,5,7,8-Pentamethylchroman-6-Sulfonyl-L-Arginine. Tetrahedron 47:6353–6370

36. Albericio F, Nicolas E, Rizo J et al (1990) convenient synthesis of fluorenylmethyl-based side-chain derivatives of glutamic and aspartic acids, lysine, and cysteine. Synthesis-Stuttgart 119–122
37. Fields CG, Lovdahl CM, Miles AJ et al (1993) Solid-phase synthesis and stability of triple-helical peptides incorporating native collagen sequences. Biopolymers 33:1695–1707
38. Low DW, Hill MG, Carrasco MR et al (2001) Total synthesis of cytochrome b562 by native chemical ligation using a removable auxiliary. P Natl Acad Sci USA 98:6554–6559
39. Tam JP, Miao ZW (1999) Stereospecific pseudoproline ligation of N-terminal serine, threonine, or cysteine-containing unprotected peptides. J Am Chem Soc 121:9013–9022
40. Hu L, Xu SL, Zhao ZG et al (2016) Ynamides as racemization-free coupling reagents for amide and peptide synthesis. J Am Chem Soc 138:13135–13138
41. Nguyen GKT, Hemu X, Quek JP et al (2016) Butelase-mediated macrocyclization of d-amino-acid-containing peptides. Angewandte Chemie-Int Edn 55:12802–12806

Chapter 5
Characterization of Triple Helix Stability Through Circular Dichroism Spectroscopy

Abstract Circular dichroism (CD) spectroscopy serves as an indispensable tool for probing biomolecular conformation and function, offering insights that are often elusive to conventional spectroscopic methods like fluorescence and absorbance. It enables rapid assessment of protein secondary structure and detection of conformational changes, even with minimal protein quantities, owing to the distinctive CD spectra associated with different structural motifs. This chapter is dedicated to elucidating triple helix stability using CD spectroscopy, commencing with an overview of circular dichroism and detailed protocols for CD measurements, including buffer and sample preparation, as well as CD cuvettes. Further exploration involves typical CD measurements for collagen mimetic peptides (CMPs), which have been utilized to investigate the thermal stability of CMPs, revealing the significant influence of specific amino acids and their placement within the (Gly-X-Y)n triplets on CMP stability. Various strategies have been developed to enhance the thermostability of CMPs, such as incorporating templates, leveraging metal ion-ligand interactions, and optimizing terminal amino acids. Additionally, the chapter discusses CD characterization of different types of Gly substitutions and natural interruptions in collagen, such as Gly-Ala and Gly-Ser interruptions, providing novel insights into the underlying mechanisms of collagen diseases and function.

Keywords Circular dichroism · Triple helix stability · Secondary structure · Gly substitutions · Natural interruptions

5.1 Introduction to Circular Dichroism

Circular dichroism (CD) is the difference in absorption between left-handed and right-handed circularly polarized light. A light beam consists of mutually perpendicular electric (E) and magnetic (B) fields oscillating in the direction of propagation [1–3]. Circularly polarized light is characterized by the rotation of the electric field vector around the propagation direction, while maintaining a consistent magnitude, which manifests as a helical pattern along the direction of propagation. Left circularly

© Springer Nature Singapore Pte Ltd. 2024

J. Xiao, *Collagen Mimetic Peptides and Their Biophysical Characterization*,
Springer Series in Biomaterials Science and Engineering 20,
https://doi.org/10.1007/978-981-10-1941-8_5

polarized light rotates the electric vector counterclockwise in the helix, while right circularly polarized light rotates it clockwise. The combination of these two circular polarizations results in elliptically polarized light, where the electric vector traces out an ellipse. Molar circular dichroism (CD) quantifies the difference in absorbance between left and right circularly polarized light and is expressed in units of $\Delta\varepsilon$. It is also often reported in degrees of ellipticity, θ, which is defined as the angle whose tangent is the ratio of the minor to the major axis of the ellipse. The molar ellipticity $[\theta]$, measured in deg*cm^2/dmol, is calculated as $3298\Delta\varepsilon$ [4]. The value of $[\theta]$ can be obtained using the equation $[\theta] = 100\ \theta/cl$, where c is the molar concentration, and l is the path length in centimeters (cm).

Asymmetric molecules interact with light by selectively absorbing left and right circularly polarized light to varying degrees, resulting in different refractive indices and the generation of a CD signal. CD is essentially an absorption phenomenon, and thus the chromophores responsible for CD spectra are the same as those contributing to conventional absorption spectra. In proteins, CD peaks in the range of 310–255 nm originate from aromatic amino acids Trp, Tyr, and Phe, while CD peaks in the far-UV region (below 250 nm) arise from the side chains of several amino acids, including Tyr, Trp, Phe, His, and Met. The peptide bond, also known as the amide chromophore, exhibits strong CD absorption at 220 and 190 nm, corresponding to its $n-\pi^*$ and $\pi-\pi^*$ transitions, respectively [5].

CD spectroscopy has a wide range of applications spanning diverse fields, including small organic molecules and large biomolecules. It offers valuable insights into the conformation and function of biomolecules that are challenging to obtain using conventional spectroscopic techniques like fluorescence and absorbance. Particularly, CD serves as an excellent tool for rapidly assessing the secondary structure of proteins due to distinct CD spectra exhibited by different structural motifs (Fig. 5.1). Proteins rich in α-helical structure feature negative bands at 222 and 208 nm, along with a positive band at 193 nm [6]. Antiparallel β-pleated sheet-rich proteins display negative bands at 218 nm and positive bands at 195 nm [7]. Disordered proteins exhibit low ellipticity above 210 nm and negative peaks near 195 nm [8]. The collagen triple helix manifests a characteristic positive peak at 220 nm and a negative peak at 195 nm, while the collagen monomer demonstrates minimal ellipticity at 220 nm and a much weaker negative peak at 195 nm [3, 9–12].

CD spectroscopy is a highly sensitive technique that allows for the investigation of conformational changes and the characterization of protein structures using small quantities of proteins under physiological and diverse buffer conditions. It provides valuable insights into the secondary structure and folding properties of proteins, offering a rapid and efficient means of structural analysis [13]. CD spectroscopy has proven useful in estimating unknown protein structures, monitoring conformational changes due to mutations, temperature fluctuations, denaturation, and ligand binding. Ligand–protein interactions can generate distinctive extrinsic CD peaks, while real-time monitoring of protein unfolding is possible through CD spectral changes induced by temperature or denaturants. Furthermore, CD serves as a valuable tool for examining the kinetics of macromolecule-macromolecule and macromolecule-ligand interactions. However, it should be noted that CD does

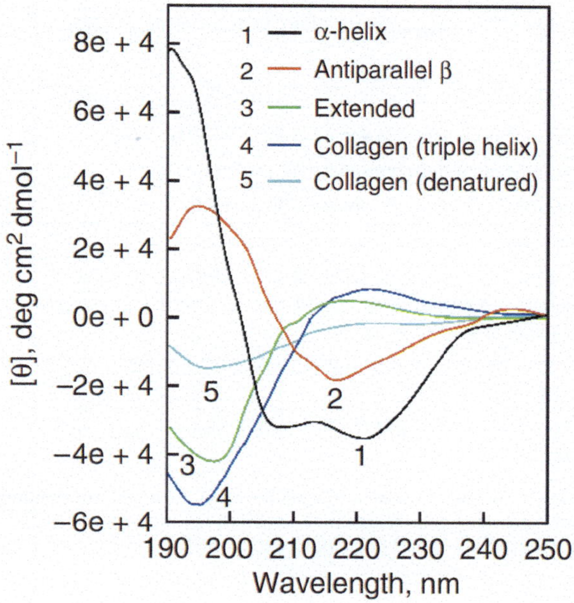

Fig. 5.1 CD spectra of proteins with representative secondary structures: α-helix (black), antiparallel β-sheet (red), disordered states (green), collagen-like triple helix (blue) and denatured collagen (cyan) [13]. Reprinted from ref. [13], Copyright © 2007, with permission from Springer Nature

not provide detailed information on the secondary structure of specific residues, as achieved by X-ray crystallography and NMR structural determinations.

5.2 Protocols for CD Measurements

5.2.1 Buffer Preparation

To ensure accurate and reliable results, CD buffers must possess certain characteristics such as optical transparency and chemical inertness. Suitable buffer options include potassium phosphate with KF, K_2SO_4, or $(NH_4)_2SO_4$ as the salt, 2 mM HEPES, and 10 mM ammonium acetate, while Tris, NaCl, and optically active components like glutamate should be avoided. Additionally, it is important to consider the cut-off value for low wavelengths, which is typically around 185 nm for commonly used phosphate buffers. It is advisable to employ distilled water for buffer preparation to reduce light absorption by oxygen. Moreover, CD buffers should be subjected to filtration through a filter with a pore size of 0.1–0.2 microns to eliminate any insoluble particles (Table 1).

Table 5.1 Buffer properties for CD experiments [13]

Buffer	~Lower wavelength limit, nm[a]
10 mM potassium phosphate, 100 mM potassium fluoride	185
10 mM potassium phosphate, 100 mM $(NH_4)_2SO_4$	185
10 mM potassium phosphate, 50 mM Na_2SO_4	185
10 mM potassium phosphate, 100 mM KCl	195
20 mM sodium phosphate, 100 mM NaCl	195
Dulbecco's phosphate buffered saline (PBS): 9.33 mM potassium phosphate, 136 mM NaCl, 2.7 mM KCl, 0.6 mM $MgCl_2$, 0.9 mM $CaCl_2$	200
2 mM HEPES, 50 mM NaCl, 2 mM EDTA, 1 mM dithiothreitol	200
50 mM Tris, 150 mM NaCl, 1 mM dithiothreitol, 0.1 mM EDTA	201

[a]The lower limit values are typically observed in solutions with protein concentrations of ~0.1 mg/mL and using cells with a path length of 0.1 cm

Reprinted from ref. [13], Copyright © 2007, with permission from Springer Nature

5.2.2 *Sample Preparation*

CD samples should be prepared using highly pure proteins or peptides with concentrations ranging from 0.005 to 5 mg/mL, depending on the path length of the CD cell. Accurate determination of protein concentrations is essential for obtaining high-quality CD data. Quantitative amino acid analysis can be employed to accurately determine protein concentration by utilizing the concentration of a selected stable amino acid. Alternatively, protein concentrations can be estimated using the measured or calculated molar extinction coefficient of the proteins [14–16].

5.2.3 *CD Cuvettes*

CD quartz cuvettes, whether rectangular or cylindrical, typically have path lengths ranging from 0.01 to 1 cm [14, 17, 18]. For rectangular cuvettes with path lengths below 0.1 cm, only a very small sample volume is required. To prevent artifacts, it is important for the CD light beam to pass directly through the sample, and in the case of cuvettes with low path lengths, the beam should be tightly focused due to their smaller surface area facing the light beam.

5.2.4 *Typical CD Measurements for Collagen Mimetic Peptides*

CD spectra are acquired using a CD spectrometer with cells having a path length of 10 mm, and the temperature of the cells is controlled using a Peltier temperature controller. The collagen mimetic peptides (CMPs) are prepared at a concentration of 1 mg/mL in 10 mM PBS at pH 7. Wavelength scans are performed from 195 to 260 nm with a 0.5 nm increment per step and a 5 s averaging time at 4 °C. Each scan is repeated three times. To assess the thermal stability, CD is employed to monitor the amplitude of the peak at 225 nm as temperature increases, with an average heating rate of 0.1 °C/min. Prior to the melting experiments, the peptides are equilibrated at 4 ºC for at least 24 h. The melting temperature (T_m) is defined as the temperature at which the folded fraction is 0.5 in the curve fitted to the trimer to monomer transition.

CD wavelength scan data are analyzed using the following steps: (1) A CD wavelength scan of the buffer is measured; (2) A CD wavelength scan of the peptide sample is measured; (3) The buffer signal is subtracted from the sample signal; (4) Any values with a dynode voltage larger than 600 are eliminated; (5) The raw CD signal (mdeg) is converted to Molecular Residue Ellipticity (MRE, deg cm^2 dmol^{-1}) using the equation: MRE = (CD_{signal} × molecular weight)/(path length × peptide bonds × concentration), where the molecular weight is in g/mol monomer, path length is in mm, peptide bonds is the number per monomer, and concentration is in mg/mL [19–21]; (6) The MRE data is smoothed; (7) Positive and negative peaks are identified, and their ratio (Rpn) is calculated, with a value of 0.12 indicating a good helical structure.

CD melting data are analyzed as follows: (1) A CD wavelength scan of the buffer is measured; (2) The CD melting data of the peptide sample is measured; (3) The CD signal of the buffer at 225 nm is subtracted from all values; (4) Linear regression fits $CD_{f(T)}$ and $CD_{u(T)}$ are obtained for the low temperature (folded) and high temperature (unfolded) linear parts of the melting curve, respectively; (5) The CD signal is converted to folded fraction (FF) using the equation: $FF_{(T)} = (CD_{(T)} - CD_{u(T)})/(CD_{f(T)} - CD_{u(T)})$, with $FF_{(T)}$ as folded fraction, $CD_{(T)}$ is the CD signal at the experimental temperature T, and $CD_{f(T)}$ and $CD_{u(T)}$ are the CD values calculated for temperature T using the linear regression equations $CD_{f(T)}$ and $CD_{u(T)}$; (6) The melting temperature (T_m) is determined when $FF_{(Tm)} = 0.5$.

5.3 CD Characterization of Collagen Mimetic Peptides

CD experiments have been utilized to investigate the thermal stability of collagen mimetic peptides (CMPs) containing the characteristic $(\text{Gly-X-Y})_n$ triplets. The stability of CMPs is significantly influenced by the specific amino acids and their placement within the $(\text{Gly-X-Y})_n$ triplets [22]. For instance, the substitution of

Proline (Pro) for Hydroxyproline (Hyp) in the Y position of the peptide (Gly-Pro-Hyp)$_{10}$ leads to a drastic decrease in the T_m from 63 to 39 °C. Moreover, interchanging the positions of Pro and Hyp results in the complete destabilization of the triple helical structure. These findings provide a fundamental understanding for the prevalence of Pro at the X position and Hyp at the Y position in the native collagen sequence.

To examine the influence of different amino acids at the X position on peptide stability, Brodsky and colleagues conducted a study using host–guest peptides Gly-X-Hyp, with sequences (GPO)$_3$-(Gly-X-Hyp)-(GPO)$_4$-Gly-Gly (Table 5.2) [23]. CD analysis of these peptides revealed that the most stable sequence was Gly-Pro-Hyp, exhibiting a T_m of 47.3 °C. Substituting Pro at the guest X position with charged amino acids led to a decrease in T_m to approximately 40.0 °C. Furthermore, the presence of a hydrophobic amino acid at the X position resulted in a significant destabilization of the triple helix structure, with the Trp-containing peptide exhibiting the lowest T_m of 31.9 °C.

To further explore the impact of different amino acids at the Y position on peptide stability, the researchers examined host–guest peptides Gly-Pro-Y, with sequences

Table 5.2 Melting temperatures of host–guest peptides with all common amino acids in the X and Y positions [23]

Gly-X-Hyp	T_m(°C)	Gly-Pro-Y	T_m(°C)
Pro	47.3	Hyp	47.3
Glu	42.9	Arg	47.2
Ala	41.7	Met	42.6
Lys	41.5	Ile	41.5
Arg	40.6	Gln	41.3
Gln	40.4	Ala	40.9
Asp	40.1	Val	40.0
Leu	39.0	Glu	39.7
Val	38.9	Thr	39.7
Met	38.6	Cys	37.7
Ile	38.4	Lys	36.8
Asn	38.3	His	35.7
Ser	38.0	Ser	35.0
His	36.5	Ala	34.0
Thr	36.2	Gly	32.7
Cys	36.1	Leu	31.7
Tyr	34.3	Asn	30.3
Phe	33.5	Tyr	30.2
Gly	33.2	Phe	28.3
Trp	31.9	Trp	26.1

Reprinted from ref. [23], Copyright © 2000, with permission from American Chemical Society

$(GPO)_3$-(Gly-Pro-Y)-$(GPO)_4$-Gly-Gly (Table 5.2) [23]. CD analysis of the Gly-Pro-Y peptides revealed that replacing Hyp in the guest Y position with other residues generally decreased the thermal stability of the triple helical structure, except for Arg. The substitution of Arg had a relatively minor effect, with a T_m value of 47.2 °C, comparable to that of Gly-Pro-Hyp (47.3 °C). In contrast, the peptide with a Trp substitution exhibited the lowest thermal stability, as indicated by a T_m value of 26.1 °C. These findings highlighted the significant influence of amino acid types at both the X and Y positions on the stability of the triple helical structure.

5.4 CD Characterization of CMPs with Stabilization Factors

The triple helical structure of CMPs plays a crucial role in determining their functionality and potential applications. Conventionally designed CMPs heavily rely on the repetitive GPO triplets for stability. CMPs with fewer than six GPO triplets typically fail to form a stable triple helix, while those containing functional motifs may face challenges in triple helix formation. Several strategies have been developed to enhance the thermostability of CMPs and overcome limitations, including the incorporation of templates, leveraging metal ion-ligand interactions, and optimizing terminal amino acids.

5.4.1 Template-Mediated Triple Helical Peptides

A number of templates have been developed to enhance the stability of triple helical structure, such as di-lysine, (KTA), and Cyclotriveratrylene (CTV) (Fig. 5.2). Heidemann et al. utilized a C-terminal covalently bridged di-lysine to assemble and stabilize a collagen-like triple helix, resulting in a T_m value less than 20 °C [24]. Goodman et al. introduced a KTA template at the N-terminal of a $(GPO)_n$ peptide and observed the formation of a triple helix with a T_m of 30 °C [25]. Liskamp et al. incorporated the chiral (+) and (−)-CTV into the N-terminus of peptide H-Gly-Gly-(Pro-Hyp-Gly)$_5$-NH_2 and demonstrated the successful stabilization of triple helices with T_m values of 55 and 58 °C, respectively. Moreover, CTV and KTA were found to effectively stabilize the triple helix of CMPs containing the blood platelet binding collagen motif GFOGERGVE, resulting in T_m values of 19 and 20 °C, respectively [26].

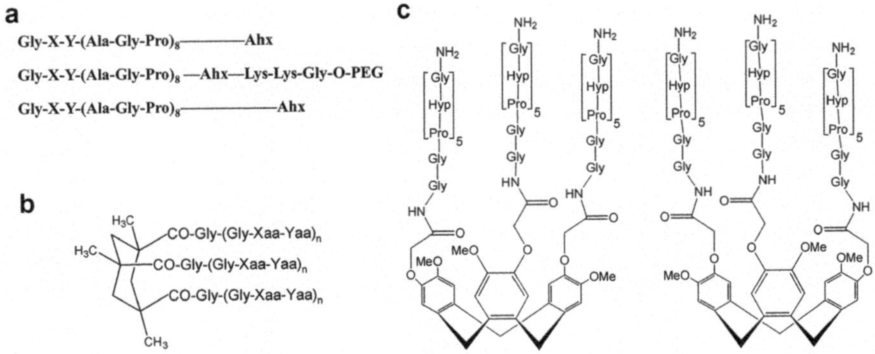

Fig. 5.2 CMPs stabilized by various templates: di-lysine (**A**) [25] (Reprinted from ref. [25], Copyright © 1998, with permission from John Wiley & Sons, Inc.), KTA (**B**) [25] (Reprinted from ref. [25], Copyright © 1998, with permission fromJohn Wiley & Sons, Inc.), and CTV (**C**) [26] (Reprinted from ref. [26], Copyright © 2002, with permission from WILEY–VCH Verlag GmbH & Co. KGaA Weinheim)

5.4.2 Metal-Assisted Assembly and Stabilization of Triple Helical Peptides

The coordination interaction between metal ions and ligands has been utilized to enhance the stability of CMPs. Koide et al. designed a peptide bpy-PP(GPP)$_7$-NH$_2$ modified with bipyridine (bpy) ligands at the N-terminal. The peptide alone showed a T_m of 33 °C, while its complex with Fe^{2+} led to a much more stable triple helix with a T_m of 46 °C (Fig. 5.3A, B) [27]. Goodman et al. introduced a dopamine residue to the N-terminal of a peptide with a (Gly-Nleu-Pro)$_6$ sequence. In the absence of Fe^{3+}, the peptide did not form a triple helix, but the presence of Fe^{3+} induced triple helix formation with a T_m of 28 °C. Additionally, they developed a peptide with a TRIS scaffold and catechols conjugated to the N-terminal and C-terminal of the (Gly-Nleu-Pro)$_6$ peptide, respectively, whose T_m was remarkably increased from 36 to 58 °C by the addition of Fe^{3+} (Fig. 5.3C, D) [28].

5.4.3 The Effect of Terminal Amino Acids on the Stability of Triple Helical Peptides

The stability of the triple helical structure was found to be influenced by the type of terminal amino acid. Wennemers et al. observed that CMPs with identical POG triplets but different terminal amino acids (P, O, or G) exhibited varying thermal stabilities [29]. CD spectra demonstrated that peptides Ac-OG(POG)$_6$-NH$_2$ and Ac-(POG)$_6$PO-NH$_2$ displayed a significant difference in T_m of 16 °C (Fig. 5.4A). Similarly, the constitutional isomers Ac-(GPO)$_7$-NH$_2$ and Ac-(POG)$_7$-NH$_2$ exhibited

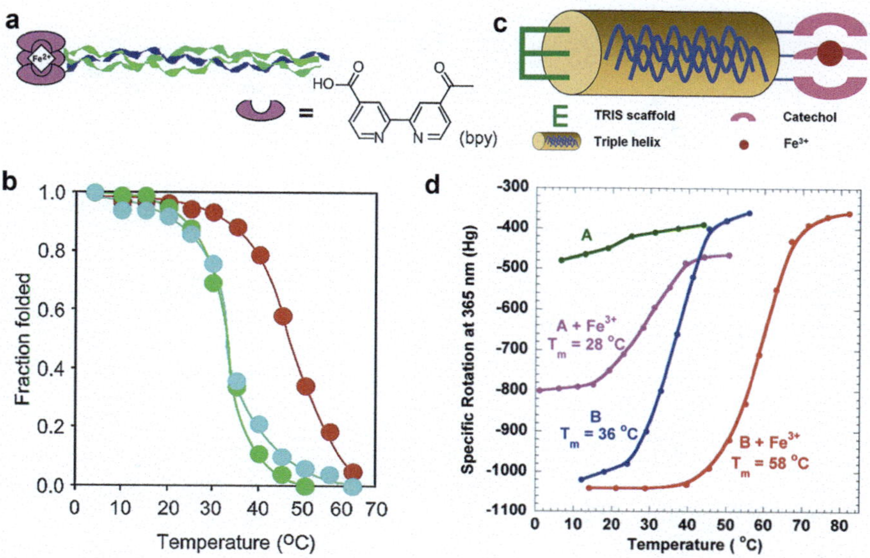

Fig. 5.3 Metal-assisted assembly and stabilization of CMPs: bpy-Fe^{2+} (**A**, **B**) [27] (Reprinted from ref. [27], Copyright © 2002, with permission from American Chemical Society), and TRIS scaffold-Fe^{3+} (**C**, **D**) [28] (Reprinted from ref. [28], Copyright © 2004, with permission from American Chemical Society)

a T_m difference of 10 °C (Fig. 5.4B). These findings indicated that the stability discrepancy arose from the preorganization of peptide termini into a polyproline-II helix.

Li et al. recently conducted a study revealing that CMPs with different terminal residues exhibited distinct end structures and thermal stabilities [30]. CD thermal unfolding curves demonstrated that the T_m value of peptide Ac-$(POG)_7$-NH_2 was 10 °C lower than that of peptide Ac-$(GPO)_7$-NH_2 (Fig. 5.4C, D). The researchers proposed that hydrogen bonds, facilitated by the terminal amino acids, played a crucial role in maintaining the end structure of the triple helix. Disruption of these hydrogen bonds led to the destabilization of the triple helix structure. These findings provide valuable strategies for modulating the stability of CMPs and hold promising applications in the fields of synthetic materials and biological probes.

5.5 CD Characterization of Different Types of Gly Substitutions

The $(Gly\text{-}X\text{-}Y)_n$ sequence pattern is absolutely conserved throughout the triple helical domain of fibrillar collagens, and even a single Gly substitution leads to osteogenesis imperfecta (OI). To systematically evaluate the effects of Gly mutations, Brodsky

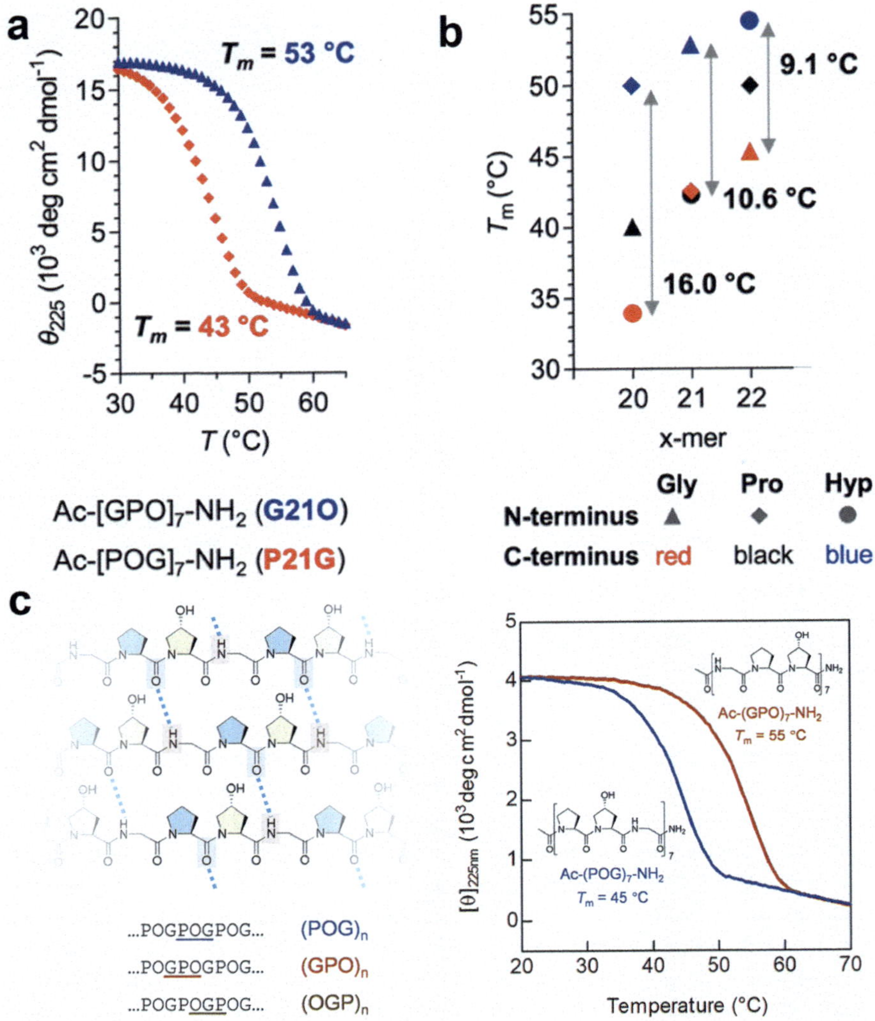

Fig. 5.4 The effect of terminal amino acids on the stability of CMPs: preorganization of peptide termini into a polyproline-II helix (**A**, **B**) [29] (Reprinted from ref. [29], Copyright © 2022, with permission from American Chemical Society), and the crucial role of hydrogen bonds (**C**, **D**) [30] (Reprinted from ref. [30], open access)

and colleagues designed a series of peptides with the sequence (Gly-Pro-Hyp)$_3$-Zaa-Pro-Hyp-(Gly-Pro-Hyp)$_4$-Gly-Gly, where Zaa represents different amino acid substitutions including Gly, Ala, Arg, Asp, Glu, Cys, Ser, or Val. The peptide GPO, with no substitutions, exhibited a stable triple helix with a T_m of 45 °C [31]. However, any Gly substitution significantly destabilized the triple helix. Gly to Ala and Ser mutations decreased the T_m to approximately 10 °C, while Gly to Asp, Glu, Val, Arg, and Cys mutations resulted in a fully unfolded conformation in an aqueous solution [31].

To assess the relative stabilities of peptides DPO, EPO, RPO, VPO, and CPO, the researchers utilized trimethylamine N-oxide (TMAO) as a cosolvent. TMAO is a naturally occurring osmolyte known for its ability to enhance protein stability. The addition of TMAO facilitated the formation of triple helices for all peptides, and the presence of 1 M TMAO was estimated to increase the T_m by 5 °C. The disruption of the triple helix conformation varied among the different Gly substitutions, following the order of Ala < Ser < CPO < Arg < Val < Glu < Asp, with the Gly-Ala mutation displaying the least perturbation [31]. The identity of the substituting residue at the Gly position played a crucial role in determining the destabilization level of the triple helical peptides, which showed a significant correlation with the lethality of Gly mutations in type I collagen associated with OI.

The impact of Gly substitutions with Ala, Ser, Asp, and Arg on the stability of the triple helix was further investigated using the specific sequence context of the T1-898 model (Table 5.3) [32]. The incorporation of an Ala mutation in the host–guest peptide T1-898 results in a destabilization of the triple helix, leading to a decrease in the melting temperature from 35 to 15.5 °C (Table 5.3). To mitigate fraying at the N-terminal caused by a Gly to Ser substitution, a new peptide, GPOT1-898[G16S], is designed with a more stabilizing $(GPO)_4$ sequence replacing the $GPO(GAO)_3$ segment. As expected, the thermal stability of GPOT1-898[G16S] is significantly increased to 28 °C compared to the T_m of 14.5 °C for T1-898[G16S] (Table 5.3).

Incorporating Ala was successful with the N-terminal $(GPO)(GAO)_3$ sequence in the presence of a C-terminal $(GPO)_4$ end, while Ser could only be incorporated with the more stabilizing $(GPO)_4$ N-terminus. Surprisingly, the addition of the N-terminal $(GPO)_4$ did not fully promote triple-helix formation when Arg was present ($T_m = 5$ °C) (Table 5.3). Modifying the adjacent triplet from GPV to GAA and the N-terminal

Table 5.3 The effects of the type of substituting residue and local sequence environment on the thermal stability of collagen OI mutation model peptides [32]

Peptide name	Peptide sequence	T_m (°C)
T1-898	GPO-GAO-GAO-GAO-GPV-GPA-GAR-GPO-GPO-GPO-GPO-GY	35.0
Gly → Ala T1-898[G16A]	GPO-GAO-GAO-GAO-GPV-APA-GAR-GPO-GPO-GPO-GPO-GY	15.5
Gly → Ser T1-898[G16S]	GPO-GAO-GAO-GAO-GPV-SPA-GAR-GPO-GPO-GPO-GPO-GY	14.5
GPOT1-898[G16S]	GPO-GPO-GPO-GPO-GPV-SPA-GAR-GPO-GPO-GPO-GPO-GY	28.0
Gly → Arg T1-898[G16R]	GPO-GAO-GAO-GAO-GPV-RPA-GAR-GPO-GPO-GPO-GPO-GY	7.0
GPOT1-898[G16R]	GPO-GPO-GPO-GPO-GPV-RPA-GAR-GPO-GPO-GPO-GPO-GY	5.0
GAAT1-898[G16R]	GPO-GPO-GPO-GAO-GAA-RPA-GAR-GPO-GPO-GPO-GPO-GY	13.0
Gly → Asp T1-898[G16D]	GPO-GAO-GAO-GAO-GPV-DPA-GAR-GPO-GPO-GPO-GPO-GY	<5.0

Reprinted from ref. [32], Copyright © 2011, with permission from Elsevier Inc.

renucleation sequence from $(GPO)_4$ to $(GPO)_3GAO$, which creates a more flexible local environment, led to a higher degree of fully folded triple-helix in the presence of a Gly → Arg mutation (T_m = 13 °C). The presence of the bulky V15 residue adjacent to Arg may hinder the required modifications for incorporating Arg, which could be facilitated by introducing a more flexible environment [32]. These findings underscore the significance of striking a balance between strong renucleation and local flexibility when accommodating bulky residues within triple helical peptides.

5.6 CD Characterization of Natural Interruptions in Collagen

The presence of natural interruptions and mutations poses seemingly conflicting challenges for the relatively uniform triple helix structure of collagen. Natural interruptions, commonly observed in non-fibrillar collagens, deviate from the $(Gly\text{-}X\text{-}Y)_n$ amino acid sequence pattern, while any Gly substitutions in fibrillar collagens lead to pathological conditions (Fig. 5.5) [33]. Connective tissue disorders have been associated with over 700 different Gly mutations in type I collagen [34, 35]. Conversely, more than 350 natural interruptions in the $(Gly\text{-}X\text{-}Y)_n$ pattern have been documented, and they are believed to have significant roles in molecular structures, collagen binding, and degradation [36]. The underlying molecular mechanisms that explain how both Gly substitutions and natural interruptions disrupt the $(Gly\text{-}X\text{-}Y)_n$ sequence pattern but yield contrasting consequences remain elusive.

Peptides are designed to mimic natural interruptions and pathological Gly substitutions within a similar sequence environment, allowing for meaningful comparisons. The interruptions are classified based on the number of amino acids between the repeating Gly-X-Y tripeptides. For example, interruptions in the form of Gly-X–Y-Gly-$(AA)_n$-Gly-X-Y are categorized as GnG type interruptions. Among these interruptions, the G5G interruptions (Gly-AA_1-AA_2-AA_3-AA_4-AA_5-Gly) are of particular interest, as they resemble the sequence pattern observed in Gly mutations

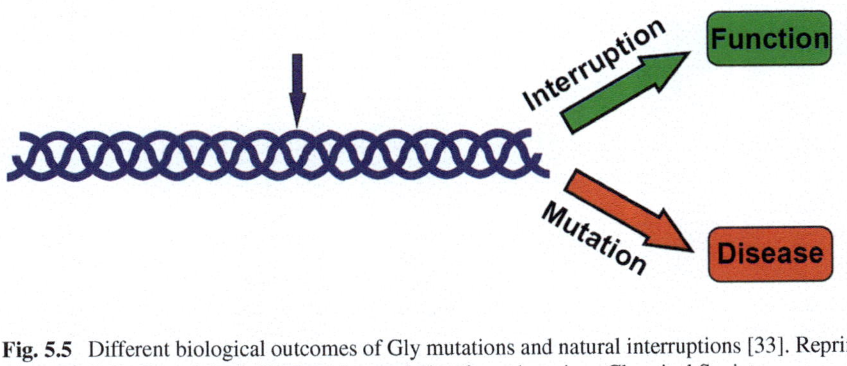

Fig. 5.5 Different biological outcomes of Gly mutations and natural interruptions [33]. Reprinted from ref. [33], Copyright © 2015, with permission from American Chemical Society

within the collagen triple helix domain, where AA_3 represents the mutant residue (Gly-AA_1-AA_2-Mut-AA_4-AA_5-Gly) [33].

5.6.1 CD Characterization of a Gly-Ala Interruption

With the aid of sequence analysis, two peptides were designed to mimic a natural G5G interruption (G-POALO-G) and a pathological Gly-Ala substitution (G-LOAPO-G). Despite the similarities in their sequences, the triple helices formed by the POALO peptide exhibit significantly higher stability compared to those formed by the LOAPO peptide. Peptide POALO (with sequence $(GPO)_5GPOALO(GPO)_4$) displays a high molecular residue ellipticity signal at 225 nm ($MRE_{225} = 4588$ deg cm^2 dmol^{-1}), and undergoes a sharp thermal transition with a T_m of 55 °C (Fig. 5.6) [33]. In contrast, the LOAPO peptide (sequence: $(GPO)_5GLOAPO(GPO)_4$) exhibits a considerably reduced MRE value ($MRE_{225} = 2370$ deg cm^2 dmol^{-1}) and does not display a well-defined melting curve (Fig. 5.6). These findings suggest that altering the relative positions of PO and LO around the Ala residue results in a significant loss of triple-helix structure. The higher stability observed in the natural interruption sequence, as compared to the Gly substitution sequence, provides insights into the distinct biological roles of natural interruptions and Gly substitutions.

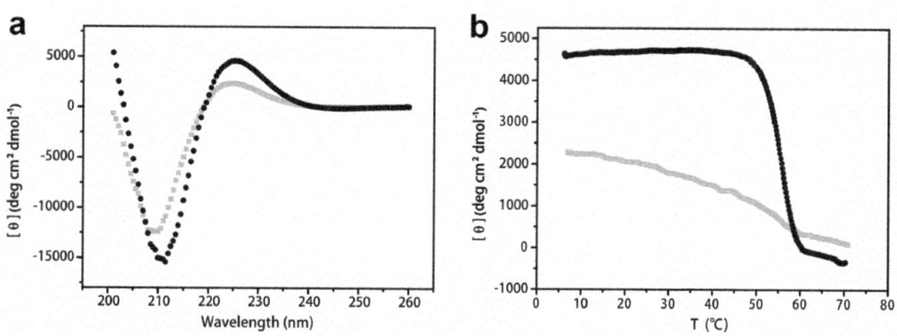

Fig. 5.6 CD wavelength scans (**A**) and thermal transition (**B**) of peptides POALO (black) and LOAPO (gray) [33]. Reprinted from ref. [33], Copyright © 2015, with permission from American Chemical Society

5.6.2 CD Characterization of a Gly-Ser Interruption

A Gly-Ser mutation with the sequence Gly-AA_1-AA_2-Ser-AA_4-AA_5-Gly and a G5G interruption (Gly-AA_1-AA_2-AA_3-AA_4-AA_5-Gly) with a central Ser at the AA_3 position share a similar local amino acid sequence context. To model a lethal Gly-Ser mutation of Type I collagen, a peptide ERSEQ (sequence: $(GPO)_6$-GERSEQ-$(GPO)_6$) is designed, while a peptide DRSER (sequence: $(GPO)_6$-GDRSER-$(GPO)_6$) is constructed to mimic a G5G interruption of Type XXIII collagen. The main distinction between the two peptides lies in the replacement of the charged amino acid Arg in DRSER with the polar residue Gln in ERSEQ.

The thermal stability of peptides ERSEQ and DRSER is examined by CD experiments (Fig. 5.7) [37]. At pH 7, peptide ERSEQ exhibits a high MRE signal at 225 nm ($MRE_{225} = 4221$ deg cm^2 dmol^{-1}). Monitoring the ellipticity at 225 nm with increasing temperature reveals two thermal transitions: a major transition with $T_m = 46.5$ °C and a minor transition with $T_m = 33.0$ °C. This suggests the presence of a lower stability species in addition to the fully folded triple helix structure of peptide ERSEQ. In contrast, peptide DRSER shows a single sharp thermal transition with a T_m of 46.0 °C. These findings indicate that while the Ser mutation exhibits a similar high stability as the interruption sequence at pH 7, the existence of a lower stability species may indicate challenges in the folding process of the Ser mutation.

The two peptides are also characterized at pH3 to investigate the electrostatic effect on their triple helix stability (Fig. 5.7). Peptide ERSEQ retains its triple helix conformation with a decreased MRE_{225} value of 3410 deg cm^2 dmol^{-1} and a lowered T_m of 40.8 °C, which is 5.7 °C lower than that at pH 7. These results indicate that the

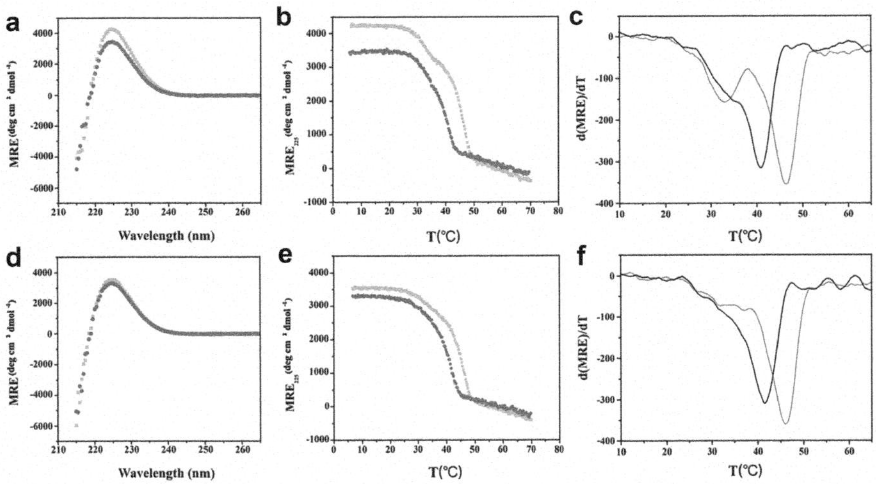

Fig. 5.7 CD wavelength scans (**A**, **D**), thermal transitions (**B**, **E**) and the d(MRE)/dT (**C**, **F**) of peptides ERSEQ (**A**–**C**) and DRSER (**D**–**F**) at pH 7 (gray) pH 3 (black) [37] (Reprinted from ref. [37], Copyright © 2015, with permission from The Protein Society)

electrostatic interaction of the charged amino acids RSE contributes significantly to the stability of the mutation site. Peptide DRSER exhibits a decreased MRE_{225} value of 3284 deg cm^2 $dmol^{-1}$ and a T_m of 41.7 °C at pH 3, indicating the same stabilizing effect of the RSE sequence at the interruption site. The slightly less decrease in stability (4.3 °C) for DRSER compared to ERSEQ (5.7 °C) upon changing pH from neutral to acidic may be attributed to the potential favorable side chain interactions involving Arg but not Gln at pH3.

References

1. Velluz L, Legrand M, Grosjean M (1965) Optical circular dichroism: principles, measurements, and applications. Verlag Chemie Academic Press Inc., New York and London
2. Sreerama N, Woody RW (2004) Computation and analysis of protein circular dichroism spectra. Methods Enzymol 383:318–351
3. Woody RW (1992) Circular dichroism and conformation of unordered polypeptides. Adv Biophys Chem 2:31–79
4. Beychok S (1966) Circular dichroism of biological macromolecules. Science 154:1288–1299
5. Yang JT, Wu CS, Martinez HM (1986) Calculation of protein conformation from circular dichroism. Method enzymol. 130:208–269
6. Holzwarth G, Doty P (1965) The ultraviolet circular dichroism of polypeptides. J Am Chem Soc 87:218–228
7. Greenfield N, Fasman GD (1969) Computed circular dichroism spectra for the evaluation of protein conformation. Biochemistry 8:4108–4116
8. Venyaminov S, Baikalov IA, Shen ZM et al (1993) Circular dichroic analysis of denatured proteins: inclusion of denatured proteins in the reference set. Anal biochem 214:17–24
9. Woody RW (2009) Circular dichroism spectrum of peptides in the Poly(Pro)II conformation. J Am Chem Soc 131:8234–8245
10. Bovey FA, Hood FP (1967) Circular dichroism spectrum of poly-L-proline. Biopolymers 5:325–326
11. Tiffany ML, Krimm S (1972) Effect of temperature on the circular dichroism spectra of polypeptides in the extended state. Biopolymers 11:2309–2316
12. Rucker AL, Creamer TP (2002) Polyproline II helical structure in protein unfolded states: lysine peptides revisited. Protein Sci 11:980–985
13. Greenfield NJ (2006) Using circular dichroism spectra to estimate protein secondary structure. Nat Protoc 1:2876–2890
14. Martin SR, Schilstra MJ (2008) Circular dichroism and its application to the study of biomolecules. Methods Cell Biol 84:263–293
15. Miles AJ, Janes RW, Wallace BA (2021) Tools and methods for circular dichroism spectroscopy of proteins: a tutorial review. Chem Soc Rev 50(15):8400–8413
16. Miles AJ, Wallace BA (2016) Circular dichroism spectroscopy of membrane proteins. Chem Soc Rev 45(18):4859–4872
17. Waldron DE, Marrington R, Grant MC et al (2010) Capillary circular dichroism. Chirality 22(1E):E136–E141
18. Daviter T, Chmel N, Rodger A (2013) Circular and linear dichroism spectroscopy for the study of protein–ligand interactions. In: Protein-ligand interactions: methods and applications, pp 211–241
19. Yang JT, Wu CS, Martinez HM (1986) Calculation of protein conformation from circular dichroism. Methods Enzymol 130:208–269
20. Greenfield NJ (1996) Methods to estimate the conformation of proteins and polypeptides from circular dichroism data. Anal Biochem 235:1–10

21. Provencher SW, Glo¨ckner J, (1981) Estimation of globular protein secondary structure from circular dichroism. Biochemistry 20:33–37
22. Cai XD, Wei WY, Bi YH et al (2020) An intrinsically nontrimerizing peptide probe for specifically targeting pathological collagen in connective tissues. Adv Funct Mater 30:2004532
23. Persikov AV, Ramshaw JA, Kirkpatrick A et al (2000) Amino acid propensities for the collagen triple-helix. Biochemistry 39:14960–14967
24. Roth W, Heidemann E (1980) Triple helix-coil transition of covalently bridged collagen-like peptides. Biopolymers 19:1909–1917
25. Goodman M, Bhumralkar M, Jefferson EA et al (1998) Collagen mimetics. Biopolymers 47:127–142
26. Rump ET, Rijkers DTS, Hilbers HW et al (2002) Cyclotriveratrylene (CTV) as a new chiral triacid scaffold capable of inducing triple helix formation of collagen peptides containing either a native sequence or Pro-Hyp-Gly repeats. Chem-Eur J 8:4613–4621
27. Koide T, Yuguchi M, Kawakita M et al (2002) Metal-assisted stabilization and probing of collagenous triple helices. J Am Chem Soc 124:9388–9389
28. Cai W, Kwok SW, Taulane JP et al (2004) Metal-assisted assembly and stabilization of collagen-like triple helices. J Am Chem Soc 126:15030–15031
29. Fiala T, Barros EP, Ebert MO et al (2022) Frame shifts affect the stability of collagen triple helices. J Am Chem Soc. https://doi.org/10.1021/jacs.2c08727
30. Qi YY, Zhou DN, Kessler JL et al (2022) Terminal repeats impact collagen triple-helix stability through hydrogen bonding. Chem Sci 13:12567–12576
31. Beck K, Chan VC, Shenoy N et al (2000) Destabilization of osteogenesis imperfecta collagen-like model peptides correlate with the identity of the residue replacing glycine. P Natl Acad Sci USA 97:4273–4278
32. Xiao JX, Madhan B, Li YJ et al (2011) Osteogenesis imperfecta model peptides: incorporation of residues replacing Gly within a triple helix achieved by renucleation and local flexibility. Biophys J 101:449–458
33. Sun X, Chai Y, Wang Q et al (2015) A natural interruption displays higher global stability and local conformational flexibility than a similar gly mutation sequence in collagen mimetic peptides. Biochemistry 54:6106–6113
34. Xiao J, Yang Z, Sun X (2015) Local amino acid sequence patterns dominate the heterogeneous phenotype for the collagen connective tissue disease Osteogenesis Imperfecta resulting from Gly mutations. J Struct Biol 192(1):127–137
35. Arseni L, Lombardi A, Orioli D (2018) From structure to phenotype: impact of collagen alterations on human health. Int J Mol Sci 19(5):1407
36. Xu T, Zhou CZ, Xiao J et al (2018) Unique conformation in a natural interruption sequence of type XIX collagen revealed by its high-resolution crystal structure. Biochemistry 57(7):1087–1095
37. Sun X, Liu S, Yu W et al (2016) CD and NMR investigation of collagen peptides mimicking a pathological Gly-Ser mutation and a natural interruption in a similar highly charged sequence context. Protein Sci 25:383–392

Chapter 6
Structural Insights from X-Ray Crystallography

Abstract X-ray crystallography, a robust experimental technique utilizing X-ray diffraction features of crystals, plays a pivotal role in determining precise molecular structures. Despite the inherent challenge of obtaining high-resolution structural information due to collagen's noncrystallizable nature, collagen mimetic peptides have transformed our understanding of collagen's biophysical properties by providing single crystals, stable triple helix structures, well-defined thermal transitions, and precise control of length and chemical composition, facilitating significant breakthroughs. Dozens of collagen mimetic peptides have been resolved with crystal structures at atomic or near-atomic resolution, confirming collagen's essential triple helical features. The crystal structure of T3-785 provides insights into sequence-dependent conformational variations in the triple-helical structure, while the integrin-binding peptide GFOGER reveals significant conformational changes between unliganded and ligand-bound forms. This chapter also explores crystal structures of Gly substitutions and natural interruptions, revealing differences in hydration and hydrogen bonding. Furthermore, it investigates crystal structures of heterotrimeric collagen mimic peptides, including both AAB-type and ABC-type heterotrimers, enriching our comprehension of collagen's intricate architecture.

Keywords X-ray crystallography · Crystal structures · Collagen mimetic peptides · Triple helix · Conformational variations

6.1 Introduction to X-Ray Crystallography

X-ray crystallography is a powerful experimental technique utilizing the X-ray diffraction features of crystals to determine the precise molecular structure [1, 2]. X-rays, with their exceptionally short wavelength of around 1 Å, interact with the atoms in the crystal and produce a characteristic diffraction pattern [3]. This pattern arises from the constructive interference of the scattered X-rays and provides critical information about the arrangement and positions of the atoms in the crystal lattice [4, 5]. Through the analysis of the diffraction data, the electron density of the molecules

© Springer Nature Singapore Pte Ltd. 2024

J. Xiao, *Collagen Mimetic Peptides and Their Biophysical Characterization*,
Springer Series in Biomaterials Science and Engineering 20,
https://doi.org/10.1007/978-981-10-1941-8_6

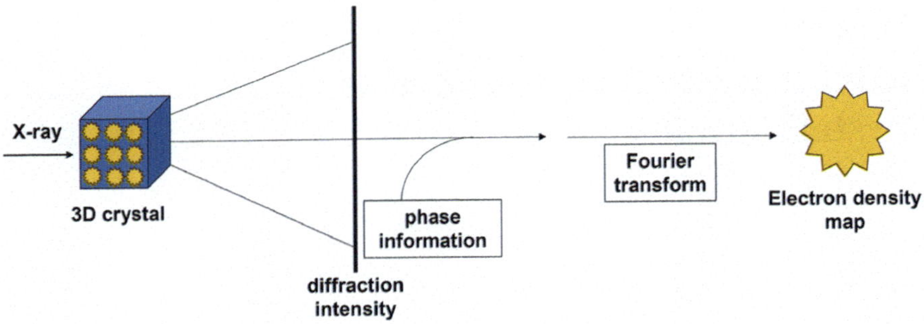

Fig. 6.1 Schematic illustration of X-ray crystallography to solve a structure

can be deduced, allowing for the construction of a detailed and accurate model of the molecular structure (Fig. 6.1) [6].

X-ray crystallography is a highly advanced biophysical technique that bears resemblance to high-resolution microscopy [7, 8]. While the resolution of microscopy is constrained by the wavelength of the electromagnetic radiation employed, X-ray crystallography surpasses these limitations [9]. Light microscopy, with a wavelength of approximately 300 nm, facilitates the visualization of cells and sub-cellular organelles [10]. Electron microscopy, on the other hand, achieves higher resolution with a wavelength as low as 10 nm, enabling the examination of intricate cellular architecture and the shapes of large proteins [11]. In contrast, X-ray crystallography exploits X-ray beams with a wavelength of approximately 1 Å, allowing for the characterization of protein structures at the atomic level.

A comprehensive understanding of protein structures is crucial for unraveling the mechanisms underlying protein functions and advancing drug development [12]. X-ray crystallography plays a pivotal role in visualizing the interaction mode of proteins with potential drugs, as well as elucidating the catalytic mechanisms of enzymes and other important biological processes [13–15]. X-ray crystallography offers a unique advantage over nuclear magnetic resonance (NMR) since it is not restricted by the size of the molecules being studied [16]. This flexibility enables X-ray crystallography to analyze a diverse range of molecular structures, spanning from small compounds to large proteins, and provides valuable insights into their three-dimensional arrangements and interactions [17–20].

In X-ray crystallography studies, obtaining a pure sample of the target proteins is a crucial prerequisite [21]. Several criteria must be met to ensure the quality of the protein sample: (1) Purity assessment using electrophoretic methods or mass spectrometry to verify the absence of impurities; (2) Confirmation of the secondary structure through circular dichroism to ensure the protein's folding integrity; (3) Evaluation of dispersity using dynamic light scattering to assess the homogeneity of the protein sample; (4) Activity testing to verify that the protein retains its functional properties; (5) Stability testing to determine the robustness of the protein under various conditions. It may be necessary to optimize expression and purification protocols, including the use of different expression systems or truncated constructs, to obtain high-quality protein samples suitable for X-ray crystallography studies.

The availability of a well-formed crystal of the target protein is a crucial requirement for obtaining accurate X-ray crystallographic structures [22]. The crystal's highly ordered three-dimensional arrangement of molecules is essential to amplify the weak diffraction signal from a single molecule [23]. Achieving exceptional crystal order allows for high-resolution diffraction measurements, leading to the determination of a high-quality protein structure [24]. Tissue culture trays are commonly utilized for setting up crystallization experiments using the hanging drop vapor diffusion method. Various compounds and temperatures can be employed to optimize crystallization conditions [25–27]. Upon identification of one or more initial hits during screening, follow-up experiments are conducted to fine-tune variables such as component concentrations, pH, buffers, and additives in order to obtain large, single crystals suitable for X-ray crystallography.

Once the crystals have been obtained, they undergo X-ray analysis, where they are rapidly cooled to a temperature of around 100 K and securely mounted on a goniometer head to allow for precise positioning in the X-ray beam [28]. Focused X-rays are then directed onto the crystal, leading to the generation of a diffraction pattern with a regular lattice of spots, indicating the presence of a single crystal. To ensure optimal data collection, various parameters such as crystal symmetry, unit cell parameters, crystal orientation, and resolution limit are carefully determined and verified. High-resolution data collection is typically performed at a synchrotron facility, where the intense X-ray beams enable the acquisition of high-quality data.

In the process of determining protein structures, one of the primary steps is to establish the phase information [29]. This can be accomplished through various techniques, such as molecular replacement, where the coordinates of a similar protein are used as an initial model to align with the experimental data. By fitting the protein backbone into the electron density map, an initial model is generated. The preliminary model is then refined against the experimental data to improve the quality of the phases. This refinement involves adjusting the model to better match the observed electron density, aiming for a crystallographic R-factor below 25%. A lower R-factor indicates a higher level of agreement between the model and experimental data. Throughout the refinement process, the ultimate goal is to generate a model that accurately represents the protein's structure, accounting for the majority of the electron density map and exhibiting meaningful chemical interactions and geometry. This final model provides a comprehensive and reliable depiction of the protein's atomic coordinates, enabling a detailed understanding of its structure and function.

6.2 X-Ray Measurements of Triple Helical Peptides

Obtaining high-resolution structural information for collagen at the molecular level has long been challenging due to its noncrystallizable nature, which limited our ability to fully understand its biophysical properties. Solid-phase synthesis of collagen mimetic peptides has provided a breakthrough in uncovering the molecular details of collagen's structure and function. These synthetic peptides have demonstrated the ability to form stable triple helix structures in dilute aqueous solutions, exhibiting distinct and well-defined thermal transitions corresponding to the denaturation of the triple helices. One notable advantage of these triple helical peptides is their precise control over length and chemical composition, resulting in homogeneous molecular masses. This inherent homogeneity makes them highly suitable for generating single crystals, a critical requirement for X-ray crystallographic studies.

The availability of single crystals derived from collagen mimetic peptides has opened up new avenues to unravel the intricate molecular architecture and functional properties of collagen. The first successful crystallization of the collagen mimetic peptide $(PPG)_{10}$, followed by high-resolution X-ray diffraction data, has provided valuable atomic-level insights into its structure [30]. However, due to the highly repetitive nature of the peptide, determining the complete structure of the entire triple helix has remained challenging. Nevertheless, the introduction of a Gly-Ala mutation in the peptide $(POG)_{10}$ has disrupted the columnar arrangements observed in $(PPG)_{10}$, facilitating the determination of the full-length crystal structure [31]. Utilizing very-high-resolution crystals, the full-length structures of both peptide $(PPG)_9$ and $(POG)_{10}$ have finally been achieved, significantly advancing our understanding of their detailed molecular architectures.

To date, dozens of collagen mimetic peptides have been solved with crystal structures at atomic or near-atomic resolution (Table 6.1) [30–33]. These structures have unequivocally confirmed the essential triple helical features of collagen, as previously observed in fiber diffraction studies of tendon. Moreover, these high-resolution structures provide detailed molecular insights into key aspects such as hydrogen bonding, hydration, and helical twist, offering a deeper understanding of the conformational variability of the collagen triple helix. Additionally, these structures have shed light on the mechanism of hydroxyproline stabilization and elucidated the interactions of collagen with other biomolecules. Overall, X-ray crystallography has been instrumental in significantly advancing our understanding of collagen and its diverse biological functions.

A concise protocol for preparing crystals of collagen mimetic peptides is outlined as follows. The peptide powder is dissolved in water to obtain a concentration of 20 mg/mL. Crystal screening is performed using the sitting-drop vapor diffusion method at a temperature of 20 °C. Within approximately one week, thin, plate-like, and rod-like crystals are typically observed. Subsequently, the optimal conditions for obtaining the best crystals are determined, such as using 0.1 M HEPES at pH 7.0 with 30% v/v Jeffamine® M-600® at pH 7.0. Before data collection, the crystals are immersed in a cryoprotectant solution containing paratone oil and paraffin. Finally,

Table 6.1 List of example collagen mimetic peptides with high-resolution crystal structures

Peptide	Sequence	Number of residues	PDB ID	Resolution (Å)	Helical twist
T3-785	(POG)$_3$ITGARGLAG(POG)$_4$	30	1BKV	2.00	Variable
G-A	(POG)$_4$POA(POG)$_5$	30	1CAG	1.85	7/2
EKG	(POG)$_4$EKG(POG)$_5$	30	1QSU	1.75	7/2
Hyp-	(POG)$_4$PG(POG)$_5$	29	1EI8	2.00	7/2
PPG10	(PPG)$_{10}$	30	1K6F	1.30	7/2
GPP-foldon	(GPP)$_9$-foldon	27	1NAY	2.60	7/2
PPG9	(PPG)$_9$	27	1ITT	1.00	7/2
POG10	(POG)$_{10}$	30	1V4F	1.26	7/2
Intergrin binding peptide	(GPO)$_2$GFOGER(GPO)$_3$	21	1Q7D	2.10	Variable
	In complex with integrin	21	1DZI	1.80	Variable
GOO9	(GOO)$_9$	27	1YM8	1.55	
OOG10	(OOG)$_{10}$	30	1WZB	1.50	
	(GPO)$_3$-G$_3$(S)OO$_2$-(GPO)$_4$	27	2G66	1.80	
LOG	(POG)$_4$-LOG-(POG)$_5$	30	2DRT	1.60	
LOG2	(POG)$_4$-(LOG)$_2$-(POG)$_4$	33	2DRX	1.40	
P-alloHyp-G	(PPG)$_4$-(P-alloHyp-G)-(PPG)$_4$	27	1X1K	1.10	
OOG	(PPG)$_4$-OOG-(PPG)$_4$	27	2D3H	1.22	
POG	(PPG)$_4$-POG-(PPG)$_4$	27	2D3F	1.26	

complete diffraction data sets are collected from the highest quality crystals at a temperature of 100 K.

The crystals' diffraction images are processed using Mosflm for indexing and integration, followed by data analysis with Aimless from the CCP4 program suite. To evaluate the model's quality, a random subset of the data (5%) is chosen for R-free calculation. ACORN is employed for crystal structure determination, utilizing a random-atom searching method to estimate initial phases. Based on the ACORN map, an initial model is constructed and refined using REFMAC5. Further refinement and model adjustments are performed using Coot, with the inclusion of water molecules guided by the Fo-Fc electron density map. Anisotropic B-factor refinement is incorporated in the final refinement process to ensure precise atomic positioning.

6.3 Crystal Structures of Collagen Mimetic Peptides with the $(Gly\text{-}X\text{-}Y)_n$ Sequences

6.3.1 Crystal Structure of a Classic Collagen Mimetic Peptide T3-785

The crystal structure of collagen mimetic peptide T3-785, resolved at a resolution of 2 Å, provides the first visualization of the sequence-dependent conformational variations in the triple-helical structure (Fig. 6.2a). Peptide T3-785 encompasses the sequence $(Pro\text{-}Hyp\text{-}Gly)_3$-Ile-Thr-Gly-Ala-Arg-Gly-Leu-Ala-Gly-Pro-Hyp-Gly-$(Pro\text{-}Hyp\text{-}Gly)_3$, with a central 12-residue, imino acid-poor segment representing the 785–796 sites of human type III collagen [34]. The helical twisting of T3-785 varied depending on its placement within the peptide, with the terminal regions showing a 7_2-helix with 3.5 residues per turn and the central region presenting a 10_3-helix with 3.3 residues per turn. While crystallographic structures of peptides $(PPG)_{10}$ and $(POG)_4POA(POG)_5$ revealed 7_2 helices, fiber diffraction studies of collagen indicated a 10_3 helix. The observed variation in the helical symmetry among different regions of T3-785 indicated that the triple-helical twist was contingent upon the peptide's unique sequence. The presence of imino acid-rich triplets $(POG)_n$ and $(PPG)_n$ imposed conformational restrictions, resulting in 7_5-fold symmetry, while the imino acid-poor region adopted a more relaxed tenfold symmetry. These findings suggest that native collagen may exhibit variable helical parameters, ranging from sevenfold in the C-terminal imino acid-rich nucleation site to tenfold in imino acid-poor regions.

Peptide T3-785 demonstrated the fundamental characteristics of the collagen triple helix, featuring direct interchain hydrogen bonding between NH groups of Gly and C = O groups of X-position residues, which is essential for the stability and structural integrity of the triple helical conformation (Fig. 6.2b). Additionally, it revealed a secondary hydrogen bond between the NH group of the X-residue and the C = O group of Gly in an adjacent chain, mediated by a lone water molecule, with this secondary hydrogen bonding occurring in the opposite direction to the primary direct hydrogen bonding (Fig. 6.2c). The mediating water molecule created additional repetitive hydrogen bonds, including an extra contact with the Hyp hydroxyl group in some cases, and it may also interact with the side chain of an adjacent Thr, supporting its stabilization role, similar to Hyp in specific collagen regions. In contrast to the X position residues involved in water-mediated interactions, the Y position residues are hydrated by water molecules directed into the solvent.

Peptide T3-785 illustrates two extreme triple-helical sequences, varying from imino acid-free to imino acid-saturated. The subtle differences between sevenfold and tenfold symmetric triple helices, considering the long repetitive nature of native collagen molecules, could have significant long-range effects on fibril formation and function. The observation of water-mediated interchain hydrogen bonding in T3-785 provided the first experimental demonstration of a second set of backbone hydrogen

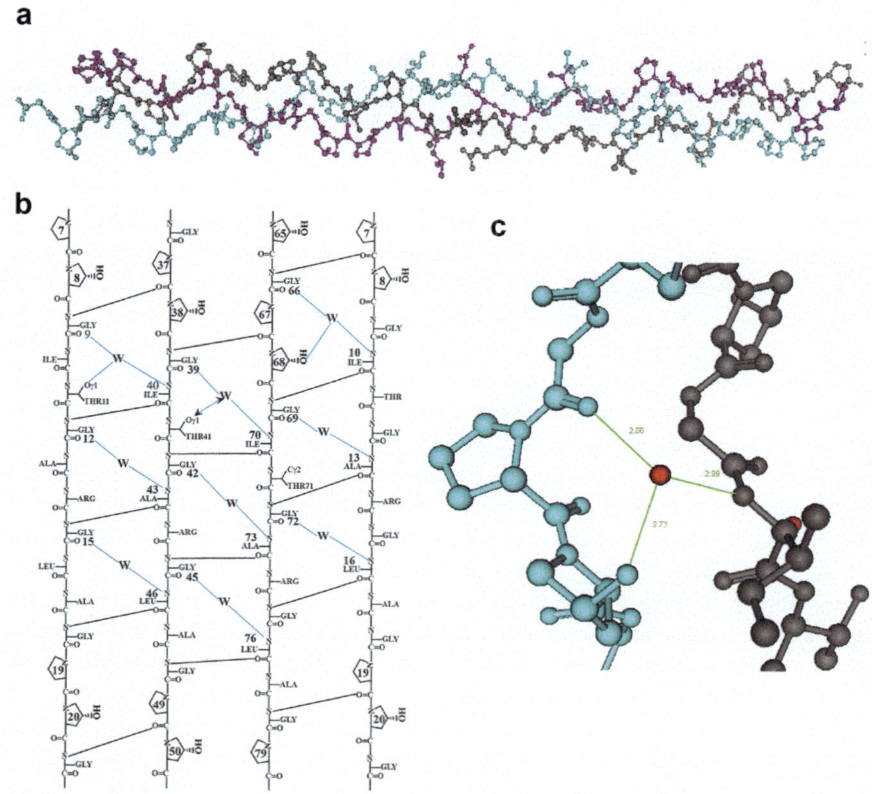

Fig. 6.2 The crystal structure of peptide T3-785 (PDB 1BKV). Stereoview (**a**) and hydrogen bonding topology (**b**) of the triple helical structure. Water molecules connect the C = O groups of Gly and NH groups from the X position residue in an adjacent chain (cyan dashed lines). Close-up view of the water-mediated interaction between Gly C = O, Ile NH and Hyp OH (**c**)

bonds in collagen, offering novel insights into how amino acid sequence influences conformational variability.

6.3.2 Crystal Structure of a Collagen Mimetic Peptide with Integrin Binding Site

The repetitive (Gly-X-Y)$_n$ peptide sequences and uniform triple helical structure of collagen present unique challenges for specific interactions with receptors and other matrix molecules. Integrin binds to collagen at multiple sites with varying affinities, and a GXX'GER binding motif has been identified as critical for this interaction. Glu is essential for specificity, as replacing it with other residues, even Asp, abolishes the binding, while Arg can be substituted by Lys or Asn, leading to reduced α_2

integrin affinity. For Type II and Type III collagen, four strong or intermediate binding sequences containing the GXX'GER motif have been identified, while 21 other sequences containing GER either weakly bind or do not bind at all, as determined by toolkit studies. Sequence analysis reveals that strong-intermediate binding sequences consistently have Hyp as X' residues, and hydrophobic or Arg residues as X residues. These distinct collagen binding affinities may dictate different biological roles, with the high-affinity GFOGER sequence capable of activating resting platelets.

The GFOGER peptide ($(GPO)_2GFOGER(GPO)_3$), designed to mimic the high-affinity integrin-binding site, exhibits significant conformational changes upon comparing its crystal structures in unliganded and ligand-bound forms (Fig. 6.3) [35]. The peptide adopts a characteristic right-handed triple helix conformation with one-residue staggering between the three chains, featuring extensive electrostatic and hydrogen bonding interactions involving the side-chains of Arg and Glu, as well as the peptide main chain, exemplified by the hydrogen bonds formed between the side-chain O-atom of Glu and the main-chain N-atom of Arg (Fig. 6.3a). The middle chain of the peptide predominantly mediated the interaction with integrin α_2 I-domain, where the Phe residue moved into a hydrophobic pocket, Glu coordinated with the critical MIDAS Mg^{2+}, and Arg formed a salt bridge with D219 (Fig. 6.3b).

The binding of integrin induces notable concerted conformational alterations in both the side chains and main chain of the GFOGER peptide (Fig. 6.3). The bound state of the GFOGER peptide showed 7_5 symmetry in its central zone, while in the free form, it displayed an intermediate degree of helical twist ranging between 7_5 and 10_7, suggesting that three imino acid-poor triplets may be the minimum requirement to trigger significant changes in helical twist away from 7_5 symmetry. Furthermore, the peptide GFOGER displays initial bending at the junction of the central region and the N- or C-terminal, and this bending was significantly enhanced following the binding interaction with the integrin α_2 I-domain [36]. Additionally, significant side-chain movements were observed in several residues during their interaction with the I domain, enabling the formation of novel hydrogen bonding. The comparison between the bound and unbound states of the GFOGER peptide reveals the inherent flexibility of the imino acid-deficient region responsible for integrin binding, which may explain the unique orientation of the three GFOGER strands in the bound structure. The extent of bending or exposure of the binding sites and specific residues to the integrin active site may play a crucial role in mediating specificity and binding affinity, providing valuable insights into the structural basis of this high-affinity interaction.

6.4 Crystal Structures of Collagen Mimetic Peptides with Gly Substitutions

The strict requirement of Gly as every third residue in the triple helical domain of fibrillar collagens is evident, as even a single Gly substitution with a bulkier residue can lead to connective tissue diseases. Peptide $(POG)_{10}$[G-A], featuring a Gly-Ala a

b

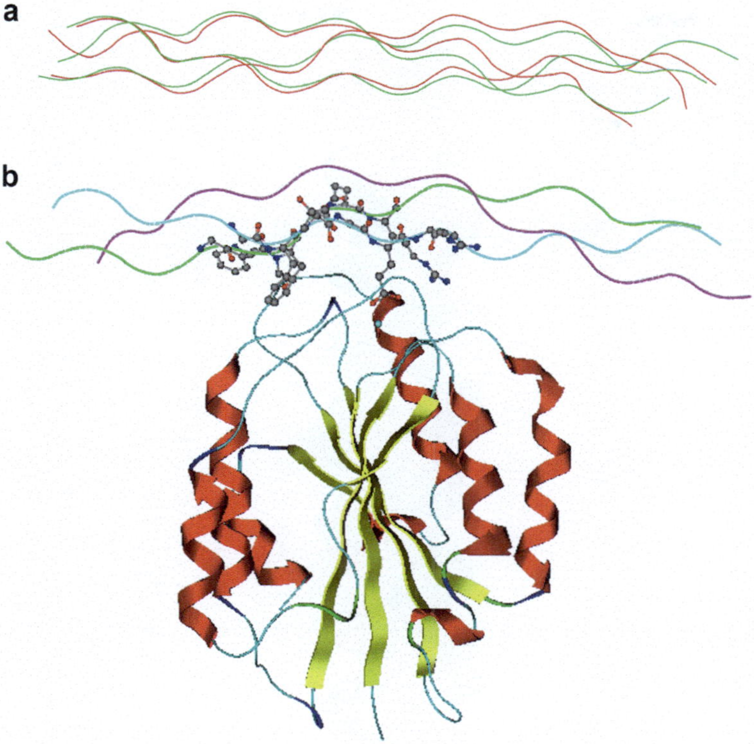

Fig. 6.3 The crystal structure of an integrin-binding peptide GFOGER (PDB 1Q7D/1ZDI). Superimposed structures of peptide GFOGER in the free and bound states (**a**); the complex structure of peptide bound with integrin α_2-I domain (**b**)

mutation in the center of $(POG)_{10}$, serves as a valuable model to investigate the effects of disrupting the repetitive Gly-X-Y triplet pattern with a single Gly substitution, aiding in the understanding of collagen Gly-X substitutions related to diseases such as Osteogenesis Imperfecta, Ehlers-Danlos syndrome IV, and Chondrodysplasia [37]. The crystals of peptide $(POG)_{10}$[G-A] displayed a straight, rod-like structure without any observable bending, contrasting with the fiber-like disorders observed in $(PPG)_{10}$ peptide (Fig. 6.4a). All residues were solvent-exposed through imino acid rings or carbonyl groups, emphasizing the critical role of solvent in stabilizing the triple helix structure. While slight perturbation was observed at the site of Ala substitution, the Ala residues remained closely packed in the triple helix without significant outward looping. In contrast, the $(POG)_{10}$[G-A] peptide displayed a twisting relaxation at the mutation site, leading to a localized unscrewing of the triple helix. The hydrogen bonding between the Gly NH and the $C = O$ group of X-position residues was maintained in the collagen zone, but around the substitution sites, steric hindrance from the methyl groups of Ala disrupted these interchain hydrogen bonds (Fig. 6.4b).

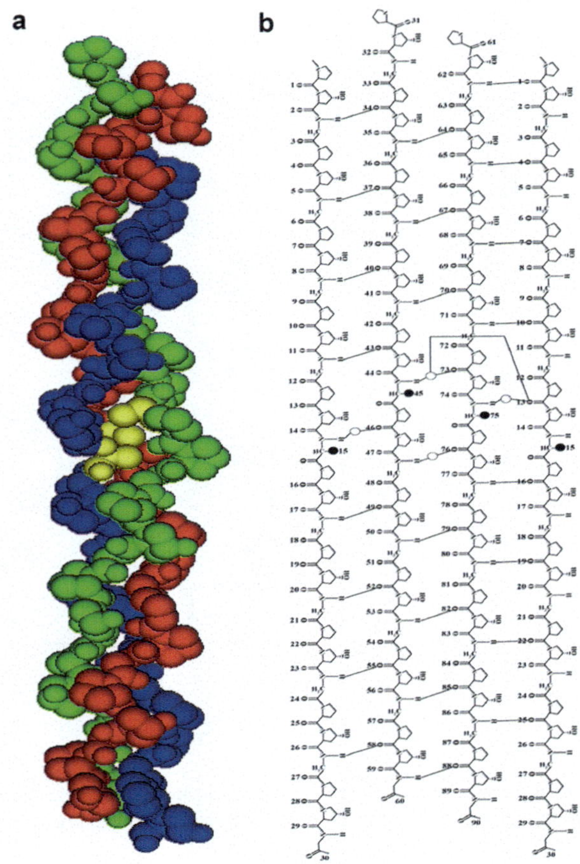

Fig. 6.4 The crystal structure of peptide $(POG)_{10}$[G-A] (PDB 1CAG). Stereoview (**a**) and hydrogen bonding topology (**b**) of the triple helical structure. Interstitial water molecules bridging the indirect hydrogen bonding at the Ala substitution site are shown as white circles

Consequently, water molecules were observed to bridge the indirect hydrogen bonds at the substitution sites.

The crystal structure of peptide $(POG)_{10}$[G-A] provided novel insights into triple helical sequences bearing Gly substitutions, as the bulkier Ala substituting residue was accommodated into the triple helix by introducing slight variations in main chain torsion angles, and water molecules effectively mediated hydrogen bonds at the substitution site. Additionally, the crystal structure showed a highly structured cylinder of hydration surrounding the triple helix, playing a crucial role in the lateral arrangement of macromolecular assemblies. Moreover, the interaction of Hyp with the hydration cylinder facilitated more hydrogen bonding opportunities for water molecules, resulting in enhanced solvation of the triple helix. The structural characteristics observed at the Ala mutation site offer valuable insights into the molecular mechanisms underlying collagen Gly substitutions and their association with connective tissue disorders.

6.5 Crystal Structures of Collagen Mimetic Peptides with Interruptions

The GPG peptide, with the sequence (Pro-Hyp-Gly)$_4$-Pro-Gly-(Pro-Hyp-Gly)$_5$, was designed to mimic a common G1G interruption found in nonfibrillar collagens, which may play crucial roles in the supramolecular organization of collagen. Peptide GPG displayed a straight triple helix without significant bending or kinking, while the interruption sites exhibited enhanced flexibility and conformational variability (Fig. 6.5a) [38]. The main-chain torsion angles in the terminal regions, enriched in imino acids, were typical of a standard triple helix with a helical symmetry close to 7_2 helix, but at the interruption site, six residues exhibited abnormal torsion angles. The terminal regions maintained standard direct hydrogen bonds between the Gly NH and the C = O group of the X-position residue, but these interactions were differently affected at the interruption site (Fig. 6.5b). The NH group of Gly14 in the leading chain faced the solvent, preventing its engagement in interchain hydrogen bonding, whereas the NH group of Gly44 in the middle chain was directed inwards, facilitating non-canonical hydrogen bonds with the C = O group of Gly72 in the lagging chain. On the other hand, the NH group of Gly74 in the lagging chain effectively participated in interchain hydrogen bonding.

The crystal structure of peptide GPG revealed a reduction in the number of interstitial ordered water molecules compared to other collagen mimetic peptides, partially due to the disruption of the water network at the interruption site. Peptide GPG showed a localized perturbation in the triple helical conformation caused by the interruption, with specific deviations in helical and superhelical parameters, torsion angles, and hydrogen bonding topology limited to a few residues around the interruption site. The interruption in peptide GPG creates a localized flexible region amidst the surrounding rigid triple helices, shedding light on the potential significance of interruption sequences in collagen function.

6.6 Crystal Structure of Heterotrimeric Collagen Mimetic Peptides

6.6.1 Crystal Structure of AAB-Type Heterotrimeric Mimetic Peptides

The design of AAB-type heterotrimeric peptides with well-defined single registers is crucial in elucidating the function of Type I collagen, as the precise positioning of the three chains is a determinant factor in mediating collagen's interactions with collagen-binding proteins (CBPs) [39]. Using a modified genetic algorithm, three heterotrimeric mimetic peptides were created, each with a specified register, where guest sequences ARGQAGVMGFO and ARGEOGNIGFO were incorporated in

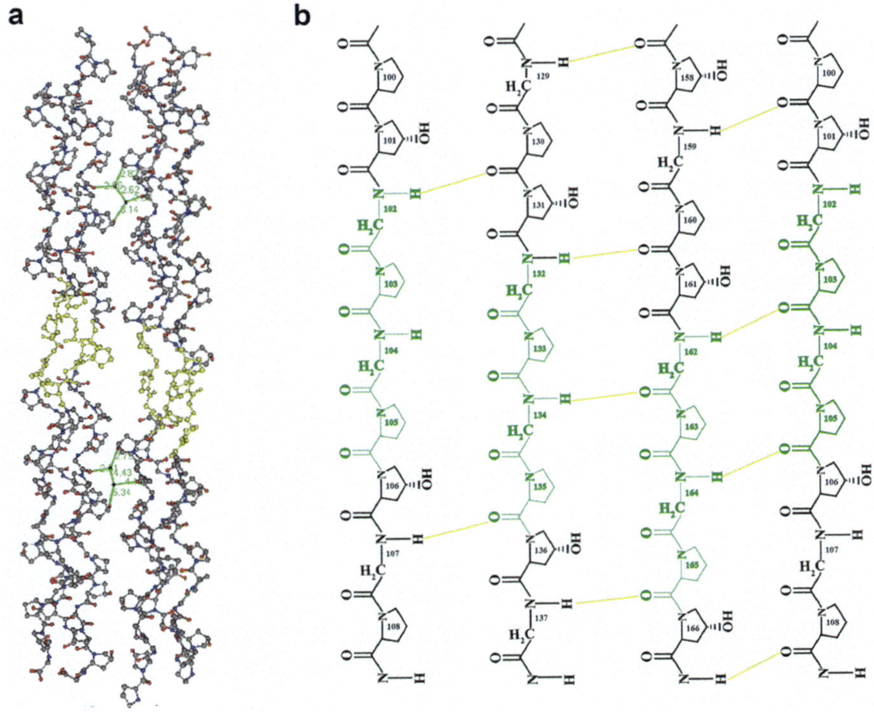

Fig. 6.5 The crystal structure of peptide GPG (PDB 1EI8). Stereoview (a) and hydrogen bonding topology (b) of the triple helical structure

chains A and B, respectively, to mimic the binding motifs of von Willebrand factor (vWF) and discoidin domain receptor (DDR) in Type I collagen (Fig. 6.6a). The heterotrimers' crystal structures have been resolved to nearly atomic resolution, confirming the accurate chain registration as designed by the computational model (Fig. 6.6b). Each peptide chain register was precisely offset by one residue relative to the adjacent chain, with chain B positioned either in the leading (BAA), middle (ABA), or trailing (AAB) position. Impressively, all three possible registers of the vWF and DDR recognition epitopes from COL1 were successfully obtained.

The crystallographic structures reveal that the homotrimeric sequence RGQOGVMGFO, which is conserved in COL2 and COL3, has a remarkable ability to recognize VWF A3 and DDR2 through the Phe residue in the leading and middle chains, respectively. Both chains A and B within the COL1 recognition sequence contain a Phe residue, enabling each CBP to potentially recognize AAB, BAA, and ABA through any of the nine feasible binding modes. Notably, the crystal structures of RGQOGVMGFO bound to DDR2 and VWF A3 accurately reproduce the interfacial interactions only when AAB binds DDR2 using the Phe residue in the middle chain and the VWF A3 domain engages with the Phe residue in the leading chain. In

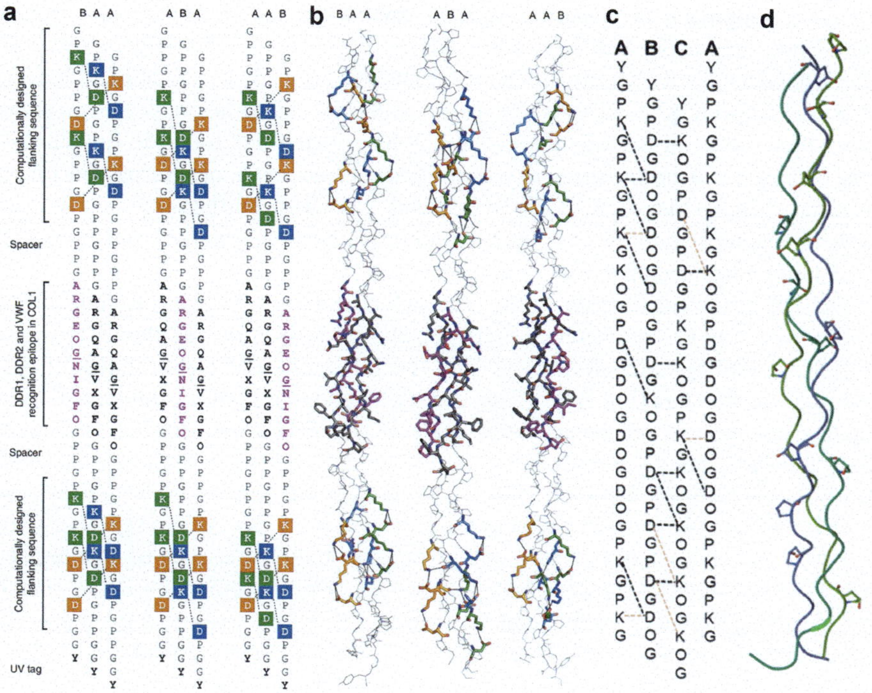

Fig. 6.6 Design of heterotrimeric registers (**a**) and corresponding crystal structures (**b**), showing the putative DDR1, DDR2 and VWF recognition epitope in COL1 (in bold) and the computationally designed flanking sequences that drive self-assembly of peptides into BAA, ABA or AAB registers. Chains A and B of the putative COL1 recognition epitope are colored black and magenta, respectively [39] (Reprinted from ref [39], Copyright © 2002, with permission from Springer Nature). For clarity, hydrogen atoms and solvent molecules are not shown. Lys–Asp salt bridges between the leading-middle, middle-trailing and trailing-leading chains are colored green, blue and orange, respectively. Sequence-based design (**c**) and atomic-resolution structure (**d**) of ABC (PDB ID code 5YAN)

contrast, the other binding modes result in the loss of one or more interfacial interactions. Therefore, heterotrimeric collagens have significantly evolved to achieve a notable level of interaction specificity with CBPs through moderate alterations in the amino acid composition and alignment of peptide chains within the triple helix.

6.6.2 *Crystal Structure of ABC-Type Heterotrimeric Collagen Mimetic Peptides*

The development of ABC-type heterotrimeric peptides is a critical advancement for elucidating the structure and mechanics of Type IV collagen and other heterotrimeric collagens. The Xu group has devised a computationally optimized salt-bridge approach to create a single-register ABC-type collagen heterotrimer [40]. Three peptides, A (PKGPKGPKGKOGPDGDOGDOGDOGPKGPKG), B (PDGDOGDOGDOGPDGKOGPDGPDGPDGDOG), and C (KOGPDGPDGPKGKOGPKGKOGKOGKOGKOG), were designed to enable favorable electrostatic interactions between peptide chains. The ABC register was found to exhibit the most extensive salt-bridge network with the least charge repulsions, surpassing other potential registers like ACB, BAC, BCA, CAB, and CBA. Furthermore, to enhance the heterotrimer's stability, two extra amino acids, tyrosine-glycine, were added at the N terminus of each peptide chain (Fig. 6.6c).

The 1.77 Å crystal structure of the ABC heterotrimer revealed a well-defined 10/3 helix with a twist parameter of $-107.6°$, closely approximating the ideal value of 108°, providing valuable insights into the intermolecular electrostatics of the triple-helical structure (Fig. 6.6d). Consistent with previous high-resolution collagen peptide structures, a recurring water-backbone side-chain network was observed along the triple helix's surface. The intermolecular interactions between adjacent chains within the triple-helical architecture occurred both parallel to the helical axis (Y–X') and radially, perpendicular to the helical axis (Y–X). This resulted in four possible combinations of Lys and Asp, giving rise to eight varieties of charge-pair interactions. A total of 21 interchain salt bridges were detected between Asp and Lys, involving both axial and lateral interactions. The inter-residue charged groups in the collagen triple-helical structure showed weaker strength for axial DK (7.05 ± 1.43 Å) and lateral KD (6.3 ± 0.62 Å) interactions, while axial KD (2.9 ± 0.19 Å) and lateral DK (4.6 ± 0.76 Å) interactions consistently demonstrated greater stability and strength. The in-depth structural analysis sheds light on the intermolecular electrostatic interactions that govern the stability and function of the ABC heterotrimer, significantly enhancing our understanding of collagen's complex architecture.

References

1. Srivastava A, Nagai T, Srivastava A et al (2018) Role of computational methods in going beyond X-ray crystallography to explore protein structure and dynamics. IJMS 19:3401
2. Papageorgiou AC, Poudel N, Mattsson J (2021) Protein structure analysis and validation with X-ray crystallography. Methods Mol Biol 2178:377–404
3. Miller RJ (2014) Femtosecond crystallography with ultrabright electrons and x-rays: capturing chemistry in action. Science 343(6175):1108–1116
4. Stetefeld J, Frank S, Jenny M et al (2003) Collagen stabilization at atomic level: crystal structure of designed $(GlyProPro)_{10}$ foldon. Structure 11:339–346
5. Ali A, Chiang YW, Santos RM (2022) X-ray diffraction techniques for mineral characterization: a review for engineers of the fundamentals, applications, and research directions. Minerals 12(2):205
6. Wang H-W, Wang J-W (2017) How cryo-electron microscopy and X-ray crystallography complement each other: Cryo-EM and X-ray crystallography complement each other. Protein Sci 26:32–39
7. Hajdu J (2000) Single-molecule X-ray diffraction. Curr Opin Struct Biol 10:569–573
8. Nakasako M, Kobayashi A, Takayama Y et al (2020) Methods and application of coherent X-ray diffraction imaging of noncrystalline particles. Biophys Rev 12:541–567

9. Lyu Z, Yao L, Chen W et al (2023) Electron microscopy studies of soft nanomaterials. Chem Rev 123(7):4051–4145
10. Sun X, Mosleh N (2023) Fluorescent carbon dots for super-resolution microscopy. Materials 16(3):890
11. Dahlberg PD, Moerner WE (2021) Cryogenic super-resolution fluorescence and electron microscopy correlated at the nanoscale. Annu Rev Phys Chem 72:253–278
12. Yuan T, Werman JM, Sampson NS (2021) The pursuit of mechanism of action: uncovering drug complexity in TB drug discovery. RSC Chem Biol 2(2):423–440
13. Shashidhara TS, Navyashree CS, Hema MK et al (2023) Unveiling the crystal structure and quantum properties of 6-bromo-N-pyridin-4-yl-2-thiophen-2-ylquinoline-4-carboxamide: a promising journey towards predicting its anticancer potential. J Mol Struct 1293:136266
14. Ji C, Wei J, Zhang L et al (2023) Aptamer-protein interactions: from regulation to biomolecular detection. Chem Rev 123(22):12471–12506
15. Nakasako M (2004) Water–protein interactions from high–resolution protein crystallography. Phil Trans R Soc Lond B 359:1191–1206
16. Tsegaye S, Dedefo G, Mehdi M (2021) Biophysical applications in structural and molecular biology. Biol Chem 402(10):1155–1177
17. Saville JW, Berezuk AM, Srivastava SS et al (2022) Three-dimensional visualization of viral structure, entry, and replication underlying the spread of SARS-CoV-2. Chem Rev 122(17):14066–14084
18. Chmelova K, Gao T, Polak M et al (2023) Multimeric structure of a subfamily III haloalkane dehalogenase-like enzyme solved by combination of cryo-EM and x-ray crystallography. Protein Sci 32(10):e4751
19. Thompson AJ, Orué AIC, Nair AJ et al (2021) Elastically flexible molecular crystals. Chem Soc Rev 50(21):11725–11740
20. Li Y, Pustovalova Y, Shi W et al (2023) Crystal structure of the CoV-Y domain of SARS-CoV-2 nonstructural protein 3. Sci Rep 13(1):2890
21. Maeki M, Ito S, Takeda R et al (2020) Room-temperature crystallography using a microfluidic protein crystal array device and its application to protein–ligand complex structure analysis. Chem Sci 11(34):9072–9087
22. Billings E, Lundquist K, Overly C et al (2021) Structure determination of membrane proteins using X-ray crystallography. Struct Funct Membrane Proteins 2302:101–136
23. Saha A, Nia SS, Rodríguez JA (2022) Electron diffraction of 3D molecular crystals. Chem Rev 122(17):13883–13914
24. Brändén G, Neutze R (2021) Advances and challenges in time-resolved macromolecular crystallography. Science 373(6558):eaba0954
25. Shaddel S, Grini T, Ucar S et al (2020) Struvite crystallization by using raw seawater: Improving economics and environmental footprint while maintaining phosphorus recovery and product quality. Water Res 173:115572
26. Wang Z, Gao K, Kan Y et al (2021) The coupling and competition of crystallization and phase separation, correlating thermodynamics and kinetics in OPV morphology and performances. Nat Commun 12(1):332
27. Li Z, Najeeb MA, Alves L et al (2020) Robot-accelerated perovskite investigation and discovery. Chem Mater 32(13):5650–5663
28. Li Y, Yin C, Yang X et al (2021) A nonlinear optical switchable sulfate of ultrawide bandgap. CCS Chem 3(7):2298–2306
29. Kaufmann K, Zhu C, Rosengarten AS et al (2020) Crystal symmetry determination in electron diffraction using machine learning. Science 367(6477):564–568
30. Kramer RZ, Vitagliano L, Bella J et al (1998) X-ray crystallographic determination of a collagen-like peptide with the repeating sequence $(Pro\text{-}Pro\text{-}Gly)_{11}$. J Mol Biol 280:623–638
31. Okuyama K, Hongo C, Fukushima R et al (2004) Crystal structures of collagen model peptides with Pro-Hyp-Gly repeating sequence at 1.26 Å resolution: Implications for proline ring puckering: crystal structures of collagen model peptides. Biopolymers 76:367–377

32. Kramer RZ, Venugopal MG, Bella J et al (2000) Staggered molecular packing in crystals of a collagen-like peptide with a single charged pair. J Mol Biol 301:1191–1205
33. Okuyama K (2008) Revisiting the molecular structure of collagen. Connect Tissue Res 49:299–310
34. Kramer RZ, Bella J, Mayville P et al (1999) Sequence dependent conformational variations of collagen triple-helical structure. Nat Struct Mol Biol 6:454–457
35. Emsley J, Knight G, Farndale R (2004) Structure of the integrin a2b1 binding collagen peptide. Acta Crystallogr Found Crystallogr 60:s164–s164
36. Emsley J, Knight CG, Farndale RW et al (2000) Structural basis of collagen recognition by integrin $\alpha_2\beta_1$. Cell 101:47–56
37. Bella J, Eaton M, Brodsky B, Berman HM (1994) Crystal and molecular structure of a collagen-like peptide at 1.9 A resolution. Science 266:75–81
38. Bella J, Liu J, Kramer R et al (2006) Conformational effects of Gly–X–Gly interruptions in the collagen triple Helix. J Mol Biol 362:298–311
39. Jalan AA, Sammon D, Hartgerink JD et al (2020) Chain alignment of collagen I deciphered using computationally designed heterotrimers. Nat Chem Biol 16:423–429
40. Zheng H, Lu C, Lan J et al (2018) How electrostatic networks modulate specificity and stability of collagen. Proc Natl Acad Sci USA 115:6207–6212

Chapter 7
Conformational and Dynamic Characterization of Collagen Mimic Peptides by NMR Spectroscopy

Abstract Nuclear Magnetic Resonance (NMR) is a potent tool for exploring protein structure and dynamics, yet its application to collagen is impeded by the molecule's considerable size and rod-like configuration. To overcome this challenge, $^{15}N/^{13}C$-isotopically selectively labeled triple helical peptides have been constructed to elucidate collagen's structural and functional intricacies. This chapter provides an in-depth examination of NMR techniques tailored to collagen mimetic peptides, encompassing a range of experiments such as triple resonance, NOESY, TOCSY, HNHA, ^{15}N relaxation, amide proton temperature gradients, residual dipolar coupling, and diffusion studies. The characterization of the classic collagen mimetic peptide T3-785 serves as a case study, demonstrating how NMR data, including sequential and chain assignments, inter-chain NOEs, hydrogen bonding patterns, and backbone conformations, can be leveraged to derive model structures via molecular modeling techniques. Additionally, this chapter delves into NMR investigations of homotrimeric collagen mimics, focusing on collagenase cleavage sites, as well as various substitutions, and extends to heterotrimeric collagen mimics, encompassing different types of heterotrimers and peptides modeling natural interruption sites, thereby providing a comprehensive understanding of collagen's structural diversity and functional implications.

Keywords NMR spectroscopy · Dynamics · Selective labeling · Model structures · NMR restraints

7.1 Introduction to NMR and Its Application in Collagen Mimetic Peptides

Nuclear magnetic resonance (NMR) has emerged as a potent technique for probing protein structure and dynamics, allowing the examination of small to medium-sized proteins in solution at the atomic level (Fig. 7.1) [1–3]. Unlike X-ray crystallography, NMR offers the distinct advantage of elucidating dynamic properties of biomolecules across a broad range of timescales, from picoseconds to seconds [4]. Measurements

© Springer Nature Singapore Pte Ltd. 2024

J. Xiao, *Collagen Mimetic Peptides and Their Biophysical Characterization*,
Springer Series in Biomaterials Science and Engineering 20,
https://doi.org/10.1007/978-981-10-1941-8_7

of ^{15}N relaxation and derived order parameter S2 provide valuable insights into fast, picosecond-nanosecond timescale motions, such as bond libration, vibration, and sidechain rotation [5, 6]. Conversely, hydrogen exchange experiments explore slower events occurring on the second timescale, such as local unfolding and helix breathing [7, 8]. Moreover, alterations in the exchange rates obtained from hydrogen exchange measurements can identify residues with significant dynamic fluctuations, potentially involved in interactions with other proteins [9]. This distinctive capability of NMR is crucial for gaining a deeper understanding of the molecular basis underlying diverse biological processes [10, 11].

NMR structure determination for normal globular proteins has become a standard multistep process, typically initiated with resonance assignments through pairs of triple-resonance experiments [12]. Conformational constraints, including proton-proton distances from NOEs and torsion angle constraints from coupling constants, are then derived [13]. Upon generating a starting structure based on the known amino acid sequence, a set of structures can be calculated by integrating experimentally determined NMR restraints [14]. This ensemble of structures accommodates the inherent conformational variability and dynamics of the biomolecule, allowing for a comprehensive depiction of its solution-state structure [15]. However, this traditional approach encounters notable challenges when applied to multidomain proteins and linear nucleic acids due to the insufficient number of NOEs available to accurately define their global conformation [16]. Hence, the incorporation of restraints containing long-range order information is vital for achieving precise NMR structure determination for these biomolecules [17]. Novel techniques, including residual dipolar coupling and paramagnetic enhancement, have rapidly evolved to offer crucial long-range restraints, extensively employed in NMR studies for structure refinement, validation, and determination of interdomain orientations [18, 19].

The considerable size and rod-like shape of collagen pose significant challenges for direct investigation using solution-state NMR [20]. As a more feasible alternative, researchers utilize shorter triple helical peptides to model various biologically important regions within collagen for NMR studies [21]. Despite their relative simplicity compared to collagen, triple helical peptides still entail substantial complexities for

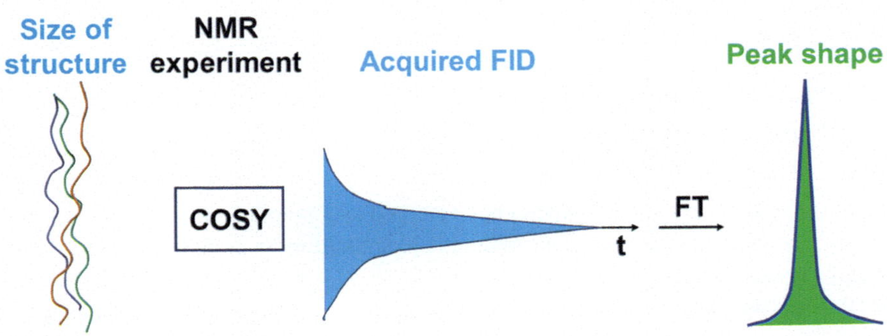

Fig. 7.1 Visualization of key features of solution NMR spectroscopy

NMR studies due to their repetitive Gly-X-Y sequences, rod-like shapes, and identical chains [22]. The rod-like structure leads to a high molecular tumbling rate, resulting in broadened peaks with decreased intensity, while the repetitive sequences and identical chains introduce significant chemical shift degeneracy [23]. Moreover, standard approaches that rely on NOEs and J-couplings to obtain structures of model triple helical peptides are limited since they typically provide only a restricted number of distances, mostly on a short-range scale [24, 25]. The overlapping resonances and lack of sufficient data render the NMR structure determination of a full triple helical peptide unfeasible.

As an alternative, researchers selectively label the triple helical peptides with ^{15}N and/or ^{13}C at specific positions, enabling the acquisition of residue-specific information through NMR measurements [26–28]. NMR studies of triple helical peptide models of collagen provide valuable insights into their native state, unfolded state, and intermediate partially folded forms [29]. These investigations have systematically explored the effects of sequence variations in collagen, including various Gly mutations and natural interruptions in the Gly-X-Y sequences, significantly advancing our understanding of the interplay between collagen sequences and their structure, dynamics, and function [30, 31].

7.2 Typical NMR Experiments for Collagen Mimetic Peptides

7.2.1 *Triple Resonance Experiments*

To achieve sequential assignments, ^{1}H–^{15}N heteronuclear single quantum coherence (HSQC) and a couple of triple resonance experiments including HNCA and HA(CA)NH/HA(CACO)NH were conducted at 10 °C. The 2D ^{1}H–^{15}N HSQC spectrum, composed of $128(t_1) \times 1024(t_2)$ complex points, was recorded using the gradient sensitivity enhanced approach, with spectral widths of 1500 and 5000 Hz in the ^{15}N and ^{1}H dimensions, respectively. The 3D HNCA experiment consisted of $80(t_1) \times 32(t_2) \times 1024(t_3)$ complex points, recorded with spectral widths of $4000(F_1)$, $1500(F_2)$ and $5000(F_3)$ Hz. The 3D HA(CA)NH and HA(CACO)NH experiments comprised $80(t_1) \times 28(t_2) \times 1024(t_3)$ and $80(t_1) \times 32(t_2) \times 1024(t_3)$ complex points, respectively, both recorded with spectral widths of $2200(F_1)$, $1600(F_2)$ and $6000(F_3)$ Hz.

7.2.2 *NOESY and TOCSY Experiments*

3D ^{15}N edited NOESY-HSQC experiments were conducted with two different mixing times, 30 ms at 20 °C and 50 ms at 10 °C. The shorter mixing time of 30 ms was used to eliminate spin diffusion, while the use of different temperatures aimed to reduce overlapped resonances. The NOESY-HSQC experiments consisted of $120(t_1) \times 32(t_2) \times 512(t_3)$ complex points, with spectral widths of 5000 Hz (F_1), 1500 Hz (F_2), and 5000 Hz (F_3). Additionally, 3D ^{15}N edited TOCSY-HSQC experiments were performed with a mixing time of 50 ms at 10 °C. The TOCSY-HSQC experiments comprised $120(t_1) \times 32(t_2) \times 512(t_3)$ complex points, and were recorded with spectral widths of $5000(F_1)$, $1500(F_2)$ and $5000(F_3)$ Hz.

7.2.3 HNHA Experiments

3D HNHA experiments were conducted with an H–H coupling period of 25 ms at 20 °C to measure $^3J_{HNH\alpha}$ coupling constants, providing information about Φ angles. The time domain data sets were composed of 512 complex points in the t_3 dimension, with 100 and 28 increments in the t_1 and t_2 dimensions, respectively. The spectral widths were set to 1600 and 6000 Hz in the ^{15}N and ^1H dimensions, respectively. A correction factor was applied to calculate $^3J_{HNH\alpha}$ coupling constants.

7.2.4 ^{15}N Relaxation Experiments

R_1, R_2 and heteronuclear NOE experiments were conducted at magnetic fields of 500, 600, and 800 MHz. For 500 MHz NMR, the spectral widths were 1500 and 5000 Hz in the ^{15}N and ^1H dimensions, respectively, with nine relaxation delays (50, 150, 250, 400, 500, 600, 700, 800, 900 ms) used in the R_1 measurement, and seven relaxation delays (4.8, 14.4, 24, 33.6, 43.2, 52.8, and 62.4 ms) employed in the R_2 measurement. For the NOE measurements, spectra were recorded in the presence and absence of ^1H saturation. Recycle delays of 1.5, 3, and 2 s were utilized in the R_1, R_2, and NOE experiments, respectively. The R_1 spectra comprised 128 (t_1) × 512 (t_2) complex points, the R_2 spectra contained 128 (t_1) × 1024 (t_2) complex points, and the NOE spectra were composed of 128 (t_1) × 512 (t_2) complex points, with 128 scans per t_1 increment used. Similar parameter setups were employed for the R_1, R_2, and NOE experiments at magnetic fields of 600 and 800 MHz. The R_1, R_2 and NOE values were calculated as the average of all data sets at each magnetic field, and the error was determined by selecting the larger value between the experimental error and the fitting error.

7.2.5 Amide Proton Temperature Gradients

To acquire hydrogen bonding information, amide proton temperature gradients were determined by conducting a series of ^1H–^{15}N HSQC spectra at temperatures ranging from 0 to 25 °C with a 5 °C interval. The sample was allowed to equilibrate for >3 h before recording each spectrum at the specified temperature. Chemical shifts of the amide protons were measured at each temperature, and the temperature gradient for each residue was calculated using linear regression of its chemical shifts against temperature.

7.2.6 Hydrogen Exchange Experiments

Hydrogen exchange experiments were performed at 10 °C and pD 2.3–2.8, with the pD corrected for the glass electrode solvent isotope artifact. Prior to the experiments, all peptide samples were equilibrated in H_2O at 10 °C for at least 48 h to ensure the monomer: trimer equilibrium was reached. Subsequently, the samples were lyophilized, redissolved in 100% D_2O, and quickly transferred to the spectrometer, which was equilibrated at 10 °C. A series of HSQC spectra were acquired every 5 min immediately after the sample was placed in the probe to measure the fast-exchanging amide protons, which were replaced by deuterons within several hours. The HSQC spectra were then acquired every 40 min to monitor the slow-exchanging amide protons. The hydrogen exchange rates (k_{ex}) were determined through non-linear least squares fitting using the equation $I(t) = (I_0 - I_\infty)\exp(-k_{ex}t)$, where I_t, I_0, and I_∞ represent the intensities at time t, at time 0 and at final time point when the exchange is complete. The protection factor P is defined as $P = k_{int}/k_{ex}$, where k_{int} is the intrinsic exchange rate of amide protons in the monomer state at a specific pH and temperature. The error of k_{int} was estimated by considering an experimental uncertainty of ±0.1 °C and ±0.02 pH units.

The hydrogen exchange process under native conditions follows a two-step model, where the amide protons in the closed state first transition into the open state, and then the open form NH becomes exposed to D and get deuterated irreversibly. The rate constant k_{ex} can be derived as $k_{ex} = k_{op}k_{int}/(k_{op} + k_{cl} + k_{int})$, where k_{op} and k_{cl} are the rates constants for structural opening and closing, respectively, and k_{int} is the intrinsic rate constant at which the amide protons in the open form, or random coil state, exchange with deuterons. When $k_{cl} \gg k_{op}$ and $k_{cl} \gg k_{int}$, k_{ex} can be simplified as $k_{ex} = k_{op}k_{int}/k_{cl} = (k_{op}/k_{cl})\, k_{int} = K_{op}\, k_{int}$. K_{op} is the equilibrium constant for structural opening and is related to the free energy for NH unfolding: $\Delta G_{HX} = -RT \ln K_{op} = -RT \ln (k_{ex}/k_{int}) = -RT \ln P$, where R is the gas constant, T is the temperature and P is the protection factor as defined above. Therefore, ΔG_{HX}, which represents the local conformational stabilities for each residue, can be calculated from the values of the protection factor P derived from hydrogen exchange measurements.

7.2.7 Residual Dipolar Coupling Experiments

To derive $^{1}D_{NH}$ (one-bond ^{15}N–^{1}H residual dipolar coupling), the difference in the corresponding J couplings ($^{1}J_{NH}$) measured in both aligned and isotropic states was calculated. The $^{1}J_{NH}$ couplings were extracted from 2D IPAP-[^{1}H, ^{15}N]-HSQC spectra, recorded as two interleaved data matrices. The differences in chemical shifts for each residue in the two matrices were determined as the values for $^{1}J_{NH}$ couplings. Polyacrylamide gels were employed to prepare the samples in the aligned state, and a 10% polyacrylamide gel was prepared following previously described protocols. After adding 30 μL 5% APS (ammonium persulfate) and 30 μL 5% TEMED (tetramethylethylenediamine) to a solution of 333 μL 30% Bis-Acrylamide and 607 μL water, the mixture was carefully pipetted into a 3 " Tygon tube (with an inner diameter of 5/32") capped with parafilm on one end. After approximately 2 min for polymerization, the gel was removed from the tube, washed for >15 h in deionized water, and cut to a length of 23 mm. The cut gel was then incubated in the peptide solution (350 μL) for >24 h at room temperature before being gently inserted into a Shigemi tube and compressed with the tube insert.

7.2.8 NMR Diffusion Experiments

Convection compensated LED followed by HSQC (CCLED-HSQC) experiments was utilized to measure residue-specific diffusion coefficients of selectively labeled peptides at 0 °C, and temperature calibration was performed using methanol at low temperatures and ethylene at high temperatures. The translational diffusion coefficients (D) were derived from non-linear regression of resonance intensity (I_t) to pulsed field gradient strength (G_t) as defined by the following equation: $I_t = I_0 * \exp\left[-2D\gamma^2 G_t^2 \delta^2 (\Delta + 2\delta/3 + 3\tau/4)\right]$, where γ is ^{1}H gyromagnetic ratio, and Δ, δ, and τ are 280, 2 and 0.4 ms, respectively. The diffusion coefficients at temperature T were then normalized to values at a selected temperature higher than the melting temperature (represented as H °C) by using the Stokes–Einstein equation. Errors in the diffusion measurements were calculated from the variance–covariance matrix, and a ratio of $D_{normalized,\,T}$ over $D_{H°C}$ was determined as the normalized diffusion coefficient for each residue, denoted as $D_{M,\,T.}$

The diffusion coefficients (D) pertain to species with the same monomer chemical shifts, which represent pure monomers at temperatures above the melting temperature, but at 0 °C, they may represent a mixture of pure monomers and partially folded species. Consequently, $D_{normalized,\,0\,°C}$ is less than or equal to $D_{H°C}$, as partially folded species have smaller diffusion coefficients due to their larger size. Therefore, the value of the normalized diffusion coefficients $D_{M0°C}$ is at most equal to 1, with a value of 1 indicating the absence of partially folded species.

7.3 NMR Characterization of a Classic Collagen Mimetic Peptide

7.3.1 *Sequential Assignments*

The triple-helical peptide T3-785, featuring the sequence $(POG)_3ITGARGLAG(POG)_4Y$, serves as a prime example of a collagen mimetic peptide that has been extensively studied through systematic NMR characterization. The HSQC spectrum of peptide T3-785 exhibited three distinct trimer resonances for all labeled residues in the central region due to non-equivalent environments among the three chains, while G24 in the repeating GPO environment displayed a single trimer resonance (Fig. 7.2a). Additionally, one monomer resonance was observed for each residue, supporting the presence of a trimeric form of the peptide in equilibrium with monomer. Residues with the same superscripted number (eg. ^{1T}G15, ^{1T}L16) in the HSQC spectrum were unambiguously assigned to the same chain. Gly had distinct chemical shifts in the ^{15}N dimension, and Gly15, being the first labeled residue in its chain, was easily recognized by its characteristic single peak in each strip. Based on the matching alpha carbon chemical shifts, three groups of sequentially connected residues were identified, represented by different colors for the corresponding peaks (red vs. black vs. green) (Fig. 7.2b). To resolve any ambiguities in the assignments from HNCA spectra, a pair of HA(CA)NH/ HA(CACO)NH experiments correlating NH in residue i to HA in residue i and/or HA in residue i–1 was performed. These results facilitated the sequential assignments of the resonances in the HSQC spectrum.

7.3.2 *Chain Assignments and Inter-Chain NOEs*

Chain assignments of peptide T3-785 are achieved using inter-chain NOEs, and all NOEs observed from the NOESY-HSQC experiments are summarized in the experimental contact map. A predicted contact map was compiled to present the information of contacts expected within 5 Å, encompassing NH–NH ((O)), NH–HA ((O)) and NH-side chain proton ((O)) contacts, with squares shaded gray for intra-chain contacts and yellow for inter-chain contacts (Fig. 7.2c). The comparison between the predicted contact map and the experimental contact map revealed a significant disparity, as only a few inter-chain NOEs, such as ^{1}G15 NH–^{2}G15 NH, ^{2}G15 NH–^{3}G15 NH, ^{1}G18 NH–^{2}G18 NH, ^{2}G18 NH–^{3}G18 NH, ^{1}L16 NH–^{3}R14 αH, ^{2}L16 NH–^{1}A17 αH and ^{3}L16 NH–^{2}A17 αH, were observed, which is much fewer than the predicted contacts (Fig. 7.2d). Inter-chain NOEs, particularly those between Gly15 in the 2nd chain with the Gly in the other two chains, support the assignment of the 2nd chain as the middle chain, while the NOE between ^{1}L16 NH–^{3}R14 αH confirms the 1st chain as the leading chain and the 3rd chain as the lagging chain, with all other inter-chain NOEs reinforcing these assignments. These inter-chain

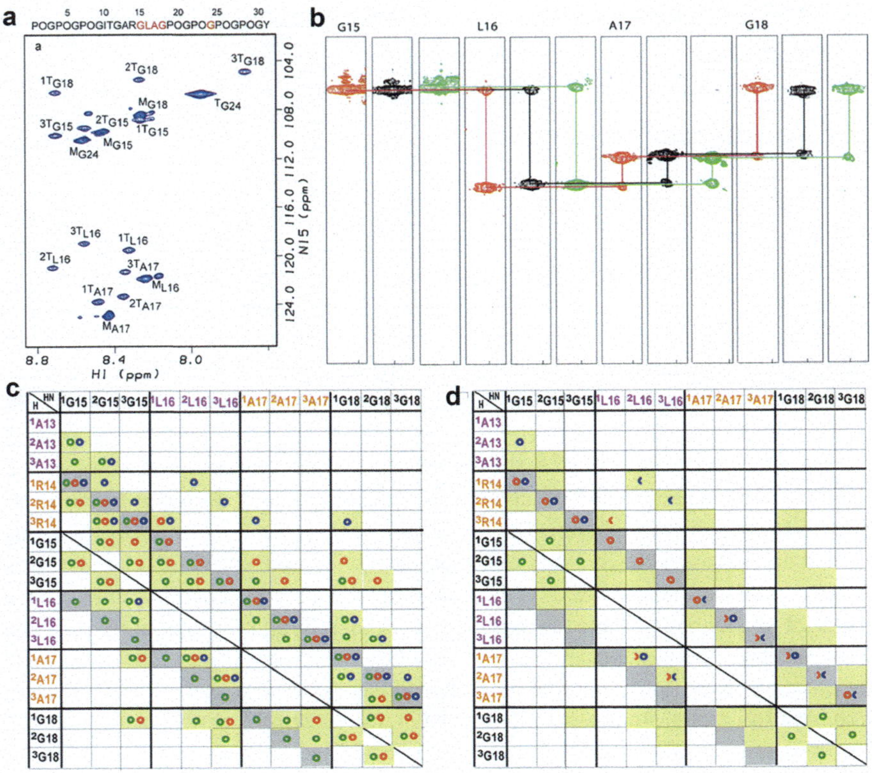

Fig. 7.2 **a** ^{1}H–^{15}N HSQC spectra of T3-785. ^{15}N/^{13}C doubly labeled residues are colored red in the sequence. The monomer and trimer peaks are denoted with a superscript M or T, respectively. The trimer resonances are further assigned to individual chains indicated by a 1, 2 or 3 referring to the leading, middle or lagging chain. **b** Strips from the HNCA spectra of T3-785. Sequential connections are shown as colored lines in individual chains. **c** Predicted contact map from the X-ray structure of T3-785. **d** Experimental contact map based on observed NOEs for T3-785

NOEs are essential for chain assignment and provide crucial distance restraints for obtaining model structures.

7.3.3 Hydrogen Bonding and Backbone Conformation

Peptide T3-785 was subjected to amide proton temperature gradient (NH $\Delta\delta/\Delta T$) measurements to investigate its hydrogen bonding patterns (Fig. 7.3a) [32]. Previous studies have indicated that amide protons with temperature gradients more positive than –4.6 are associated with the presence of hydrogen bonds. Glycine residues at positions 15, 18, and 24 displayed NH $\Delta\delta/\Delta T$ values ranging from –4 to –1, indicative of hydrogen bond formation. In contrast, residues Leu16 and Ala17 in

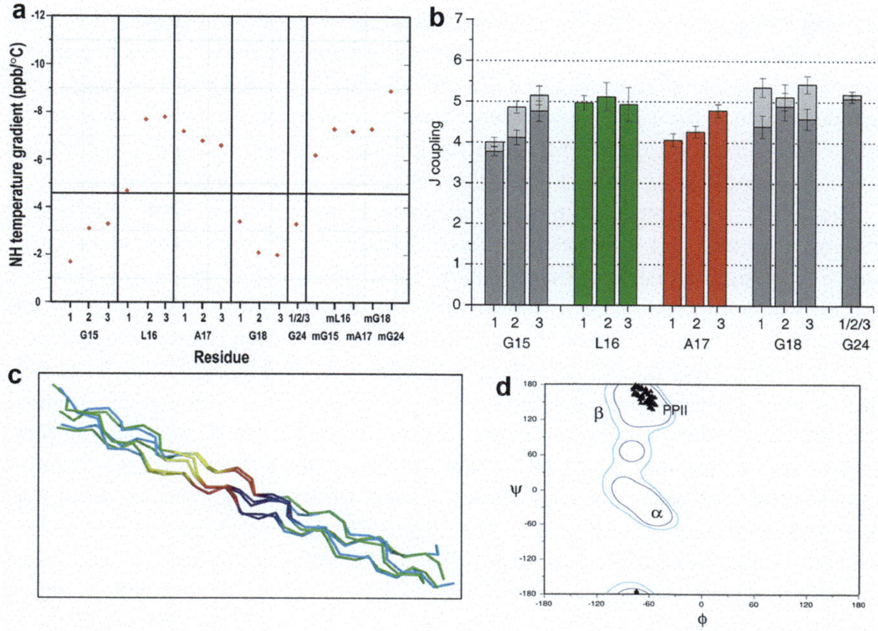

Fig. 7.3 **a** Amide proton temperature gradients ($\Delta\delta/\Delta T$) of T3-785. The black horizontal line corresponds to a value of –4.6, a cut-off for hydrogen bonding, with less negative values indicative of hydrogen bonding. **b** $^3J_{HNH\alpha}$ coupling constants of T3-785. **c** Model structure of T3-785 (cyan) aligned with its X-ray structure (green). Triplets ITG, ARG and LAG are colored in yellow, red and blue, respectively. **d** Ramachandran plot for central residues in T3-785. Favorable regions for α-helix, β-sheet and polyproline II are indicated, while favorable regions for Pro are circled

the triple helix, as well as those in the monomer state showed more negative NH $\Delta\delta/\Delta T$ values, suggesting the absence of hydrogen bonds. The dihedral angle Φ, crucial for defining backbone conformation, was determined from $^3J_{HNH\alpha}$ coupling constants, which exhibited typical values around 4 or 5, consistent with a triple helix polyproline II conformation (Fig. 7.3b) [33]. Gly exhibited two distinct measured J coupling values due to its two alpha protons.

7.3.4 Model Structure from NMR Restraints and Molecular Modeling

To overcome the challenges posed by limited NOEs in peptide T3-785 for NMR structure determination, a molecular modeling strategy that incorporated NMR restraints was employed to obtain model structures using the Molecular Operating

Environment software. The X-ray structure of the triple helical peptide $(POG)_3$-POGEKGPOG-$(POG)_4$ was used as a starting structure, and central residue mutations were introduced, replacing POGEKGPOG with ITGARGLAG, resulting in the peptide sequence for T3-785 [34]. Subsequently, energy minimization was carried out on the 9 central residues, while ensuring that the repeating POG ends maintained an idealized triple helix conformation throughout the simulations.

The NMR restraints encompassed distance restraints from NOEs, hydrogen bonding restraints, and Φ angles [35]. Inter-chain NOEs in the experimental contact map were assigned distance limits ranging from 1.8 to 4.2 Å. Hydrogen bonding restraints were imposed on Gly15 and Gly18, with distance limits set between 1.8 and 2.2 Å and N–H–O angle limits ranging from 145 to 180°. The precise Φ angle limits were determined by calculating the four solutions of the J couplings for each residue. The model structures obtained after performing energy minimization were validated using the Molprobity program to assess clashes, rotamer, and Ramachandran plot. Comparison of the validated T3-785 model structure with the X-ray structure demonstrated a close similarity, with a backbone RMSD of 1.21 Å (Fig. 7.3c). Additionally, the Ramachandran plot for the central residues of T3-785 indicated that all residues adopted the favorable triple helical polyproline II conformation (Fig. 7.3d). These results highlight the robustness of molecular modeling enriched with NMR restraints as an effective approach for obtaining precise model structures for collagen mimetic peptides.

7.4 NMR Characterization of Homotrimeric Collagen Mimetic Peptides

7.4.1 NMR Characterization of Collagen Peptides Mimicking Collagenase Cleavage Sites

The collagenase cleavage site, Gly-[Ile/Leu]-[Ala/Leu], is intriguingly present in various collagens, but several other sites with the same Gly-Ile/Leu scissile bond do not undergo hydrolysis [36]. Peptides T3-778 and T3-785 were designed to model the natural cleavage site (GPLGIAGITGAR) and a potential but non-cleavable site (GITGARGLAGPO), respectively. Additional variants T3-778I-L, T3-778P-A, and T3-778IT-PO were synthesized to explore the impact of specific residues and surrounding triplets, with T3-778I-L investigating the difference between GIA and GLA, T3-778P-A examining preceding imino acid triplets, and T3-778IT-PO studying following non-imino acid triplets. NMR HSQC spectra of all the peptides exhibited both trimer and monomer peaks for all labeled residues, indicating an equilibrium between the triple helix and monomer conformations. Peptide T3-778 showed distinct chemical shifts for the Ile in the first chain (^{1T}I17) compared to the other two chains, while peptide T3-785 exhibited closer chemical shifts for the Leu in the first chain (^{1T}L16) to the Leu in the other two chains. The three T3-778

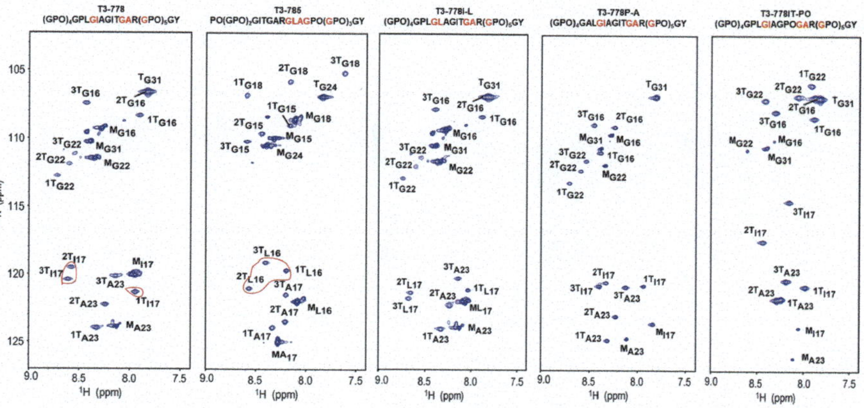

Fig. 7.4 HSQC spectra of active peptide T3-778, inactive peptide T3-785, peptide T3-778I-L, T3-778P-A and T3-778IT-PO [36]. Reprinted from ref. [36], Copyright © 2010, with permission from Elsevier Science Ltd. ^{15}N labeled residues are colored red in the above sequences. The peaks corresponding to the monomer and trimer state are denoted with a superscript M or T. The superscripted number 1, 2 and 3 correspond to the leading, middle and lagging chains, respectively

mutant peptides displayed similar HSQC spectra to T3-778, with distinct chemical shifts observed for the Ile/Leu residues in the first chain compared to the other two chains (Fig. 7.4). These findings suggested that the P-A and IT-PO mutations induced a unique chemical environment for G16 or I17 at the collagenase cleavage site, potentially influencing the specificity of collagen recognition by MMP.

J-coupling constants serve as a reliable measure of backbone conformation, particularly for residues in the triple helical conformation with phi angles ranging from –55 to –75°, which correspond to J coupling values of 4–6 Hz. Notably, all residues in the triple helix of the model peptides exhibited J-coupling values between 4 and 6 Hz, except for a single Ile in T3-778 (^{1T}I17) that displayed a value of 6.74 Hz, along with a distinct chemical shift (Fig. 7.5a). Residue-specific local stabilities, derived from hydrogen exchange experiments, revealed consistently high stabilities for all Gly residues in the peptides, while the Ala residues in the X position exhibited uniformly decreased stabilities (Fig. 7.5b). Remarkably, the Leu residues in T3-785 displayed uniform high stability, suggesting the presence of indirect hydrogen bonding, while the other Ile/Leu residues exhibited significant instability, indicating the loss of hydrogen bonding and alteration of hydration structure. Furthermore, the stabilities of Ile/Leu residues in different chains varied significantly between peptides; in T3-778 and T3-778I-L, the Ile/Leu in the first chain showed much lower stability compared to those in the other two chains, while in T3-778P-A and T3-778IT-PO, the Ile in the first chain displayed similar stabilities to the Ile in the third chain (Fig. 7.5b). The non-equivalent staggering between the three chains of T3-778, T3-778P-A, and T3-778IT-PO plays a crucial role in their distinct stabilities (Fig. 7.5c). The presence of Pro at the 3 subsites in the third chain affects only the Ile in the first chain of T3-778, while in T3-778IT-PO, an additional Ile in the third

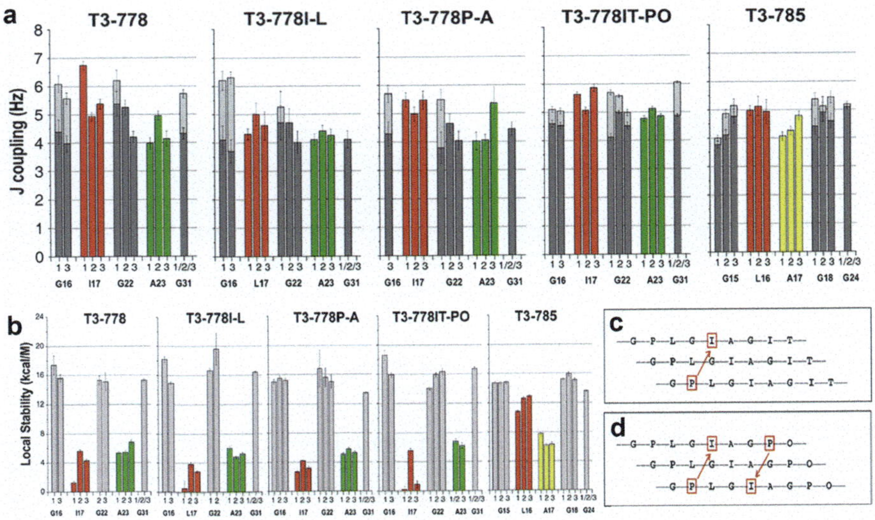

Fig. 7.5 **a** J-coupling values of model peptides. **b** Local stabilities of model peptides calculated from protection factors. Gly are colored in gray, Ile/Leu in the X position are colored in red, while Ala are colored green in the X position and yellow in the Y position. Effect of neighboring Pro on Ile at the cleavage site. **c** Pro at 4 subsite in the lagging chain affects the Ile in the leading chain; **d** Pro at 4' subsite in the leading chain affects the Ile in the lagging chain [36]. Reprinted from ref [36], Copyright © 2010, with permission from Elsevier Science Ltd.

chain is influenced by the Pro at the 4' subsite, and T3-778P-A lacks this effect due to the absence of Pro (Fig. 7.5d).

7.4.2 NMR Characterization of Gly to Ala/Ser Substitution

Peptides T1-898[G16A] and T1-898[G16S] were designed to model OI mutations at position 901 in the α_1 chain of type I collagen, and to explore the impact of Gly to Ala/Ser substitution on the equilibrium conformations of peptide T1-898 (Fig. 7.6a, e) [37]. The substitution of the central Gly with Ala retained the fully folded triple-helix characteristics, evident from the distinct trimer resonances of labeled residues at positions G7, A16, and G28 in the HSQC spectrum (Fig. 7.6b). Normalized diffusion coefficients at 0 °C ($D_{M0\,°C}$) were measured for all the labeled residues in peptide T1-898[G16A], revealing that $D_{M0\,°C}$ of G7 and G28 were close to 1, suggesting that the two terminal residues were not in partially folded states and confirming that the two ends of the peptide were folded (Fig. 7.6f). The significantly lower $D_{M0\,°C}$ value of A16 compared to G7 and G28 indicated a partially folded trimer with a break in the middle (PFT-B), indicating a looser middle region in the triple helical conformation (Fig. 7.6f).

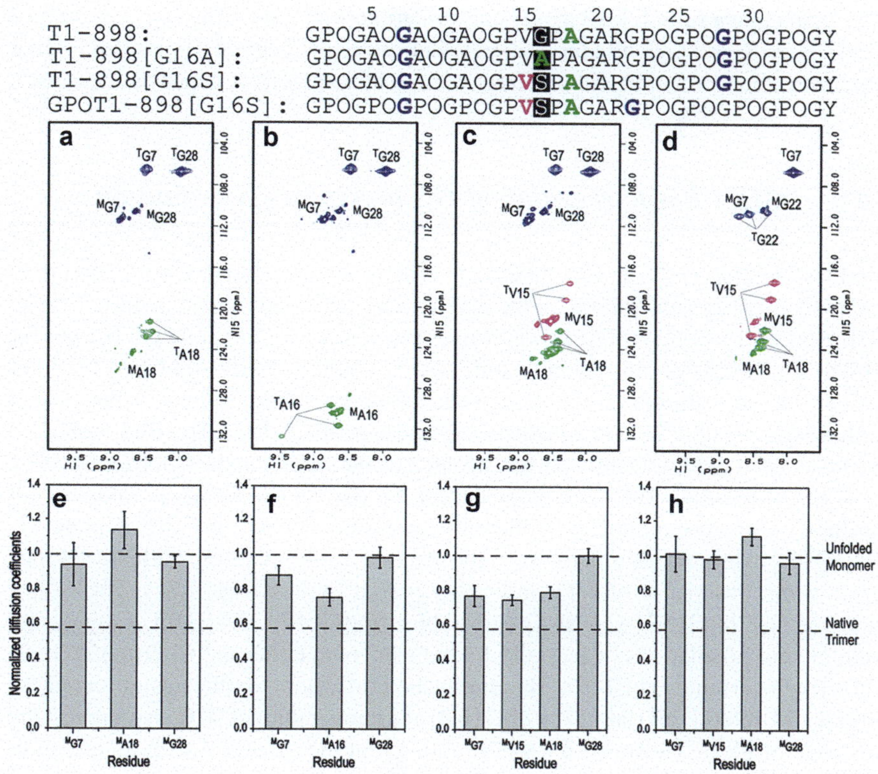

Fig. 7.6 Effect of Gly to Ala/Ser substitution on equilibrium conformations of peptide T1-898 [37] (Reprinted from ref [37], Copyright © 2011, with permission from Elsevier Science Ltd.). ^{1}H–^{15}N HSQC spectra of T1-898 (**a**), T1-898[G16A] (**b**), T1-898[G16S] (**c**) and GPOT1-898[G16S] (**d**). Histogram of the residue-specific monomer diffusion coefficients of ^{15}N-labeled residues for peptide T1-898 (**e**), T1-898[G16A] (**f**), T1-898[G16S] (**g**) and GPOT1-898[G16S] (**h**). The peaks corresponding to the monomer and trimer states are denoted with a superscript M or T, respectively. The black dashed line represents the normalized diffusion coefficients for unfolded monomer (theoretically with value of 1) and native trimer (with a value of 0.58 for G28 in peptide T1-898[G16A]), respectively

T1-898[G16S] exhibited distinct trimer resonances for labeled residues G7, V15, A18, and G28 in the HSQC spectrum, demonstrating successful incorporation of Gly to Ser substitutions into the triple helical conformation (Fig. 7.6c). The $D_{M0\,°C}$ values for G7, V15, and A18 were similar and smaller than 1, while G28 showed a value close to 1, indicating the presence of a partially folded species with a folded C-terminus but unfolded N-terminal and mutation site (PFT-C) (Fig. 7.6g). To enhance the folding of the peptide with Gly to Ser substitution, Peptide GPOT1-898[G16S] was designed by replacing the less stabilizing GPO(GAO)$_3$ at the N-terminus with a more stabilizing (GPO)$_4$ sequence. The HSQC spectrum of GPOT1-898[G16S] exhibited distinct trimer resonances for all labeled residues at G7, V15, A18, and

G22, indicating a fully folded trimer conformation (Fig. 7.6d). The $D_{M5\,°C}$ values of G7, V15, A18, and G22 all approached 1, indicating the peptide's complete folding into a trimer conformation without the presence of partially folded species (Fig. 7.6h).

7.4.3 NMR Characterization of Gly to Asp/Arg Substitution

Peptides T1-898[G16D] and T1-898[G16R] were constructed to investigate the effect of Gly to Asp/Arg substitution on the equilibrium conformations of peptide T1-898 [37]. Peptide T1-898[G16D] showed trimer resonances for G7 and G28 but not for G13 in the HSQC spectrum, indicating that it could not fully fold into a trimer through the substitution site, although the two ends might adopt a triple helical conformation (Fig. 7.7a). The diffusion measurements of T1-898[G16D] indicated the presence of predominantly partially folded trimers either at the C-terminus (PFT-C) or the N-terminus (PFT-N), as evidenced by the very low $D_{M0\,°C}$ values of all residues, closely resembling the trimer diffusion coefficients (Fig. 7.7b). Peptide T1-898[G16R] exhibited a similar HSQC spectrum to T1-898[G16D], with the absence of the trimer resonance of G13, indicating partial folding (Fig. 7.7c). The $D_{M0\,°C}$ values of T1-898[G16R] were also comparable to those of T1-898[G16D], suggesting predominant populations of partially folded trimers at either the C-terminus (PFT-C) or the N-terminus (PFT-N). The diminished diffusion coefficients observed for peptides T1-898(G16R) and T1-898(G16D) at a temperature of 0 °C also suggest the existence of PFTs. However, it's worth noting that the accuracy of these measurements might be compromised due to the overlapping resonance signals associated with the disordered structural state (Fig. 7.7g).

Peptide GPOT1-898[G16R] was designed with a similar approach to GPOT1-898[G16S], incorporating a more stabilizing $(GPO)_4$ sequence at the N-terminus in place of the less stabilizing $GPO(GAO)_3$, aiming to facilitate folding through the Gly to Arg mutation site. The HSQC spectrum of GPOT1-898[G16R] showed a weak trimer peak for residue G7 and a trimer peak for A18, while V15 next to the Gly substitution site did not show any trimer peak, indicating the presence of only partially folded species unable to fold through the mutation site. The smaller diffusion coefficients of the disordered resonances of G7 and V15 further corroborated this inference (Fig. 7.7d, h). Peptide GAAT1-898[G16R] was constructed by introducing more flexible residues, such as GAA in place of GPV next to the mutation site and $(GPO)_3GAO$ instead of $(GPO)_4$ in the N-terminal renucleation region, to enhance the incorporation of Gly substitutions. The HSQC spectrum of GAAT1-898[G16R] showed a strong trimer resonance for Gly7 and three well-dispersed trimer resonances for A15, significantly enhanced compared to GPOT1-898[Gly16R] (Fig. 7.7e). Lyophilizing and re-dissolving GAAT1-898[G16R] in 100% D_2O allowed the resolution of overlapping trimer resonances for G22, enabling the observation of trimer resonances after several hours due to faster NH exchange in the monomer. After approximately 20

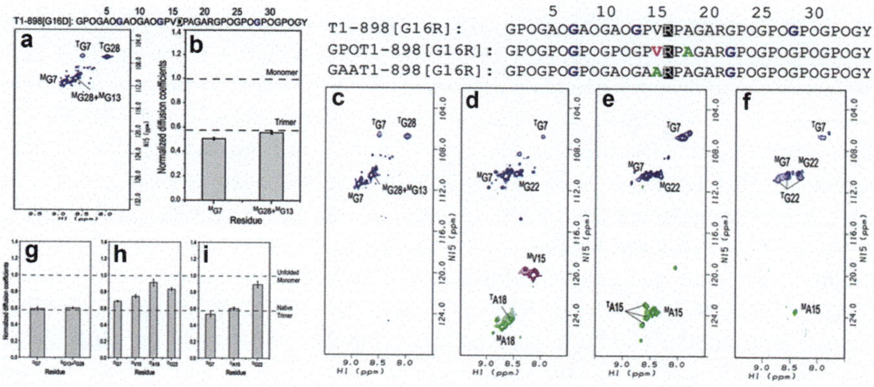

Fig. 7.7 Effect of Gly to Asp/Arg substitution. ^{1}H-^{15}N HSQC spectra of T1-898[G16D] (**a**) and histogram of the residue-specific monomer diffusion coefficients of ^{15}N-labeled residues for peptide T1-898[G16D] (**b**). The black dashed line represents the normalized diffusion coefficients for unfolded monomer (theoretically with value of 1) and native trimer (with a value of 0.58 for G28 in peptide T1-898[G16A]), respectively. ^{1}H–^{15}N HSQC spectra of T1-898[G16R] (**c**), GPOT1-898[G16R] (**d**), GAAT1-898[G16R] (**e**) and GAAT1-898[G16R] after ~20h in 100% D_2O (**f**), the peaks corresponding to the monomer and trimer states are denoted with a superscript M or T, respectively. Histograms of the residue-specific translational diffusion coefficients of ^{15}N-labeled residues in disordered states for peptide T1-898(G16R) (**g**), GPOT1-898(G16R) (**h**), and GAAT1-898(G16R) (**i**) [37]. Reprinted from ref. [37], Copyright © 2011, with permission from Elsevier Science Ltd.

h in D_2O, the HSQC spectrum of GAAT1-898[G16R] showed three trimer resonances for G22, confirming the formation of a fully folded trimer conformation, as evidenced by the trimer resonances of G7, A15, and G22 (Fig. 7.7f). These findings underscore the essential roles of local flexible environments and robust renucleation sequences in facilitating the accommodation of Asp/Arg substitutions into the triple helix. Nonetheless, it is plausible that PFTs persist within the peptide. This is suggested by the diminished diffusion coefficients observed for G7 and A15, although the accuracy of these measurements could be impeded by the presence of overlapping resonances (Fig. 7.7i).

7.5 NMR Characterization of Heterotrimeric Collagen Mimetic Peptides

7.5.1 NMR Characterization of AAB-Type Heterotrimers

Horng and co-workers explored the formation of stable AAB collagen heterotrimers through the development of cation $-\pi$ interactions between cationic and aromatic residues, utilizing two cationic peptides CR (with the sequence $(POG)_6$-$(PRG)_3$) and

CK (with the sequence $(POG)_6$-$(PKG)_3$), as well as one aromatic peptide CF(with the sequence $(POG)_6$-$(FOG)_3$) [38]. ^{1}H,^{15}N–HSQC spectra of CR and the 2CR:1CF mixture revealed the presence of new resonances in the 2CR:1CF mixture, providing evidence for the successful formation of heterotrimers consisting of CR and CF (Fig. 7.8a). The ^{1}H,^{15}N–HSQC spectrum of the 1CR:2CF mixture displays unique resonances distinct from those in the 2CR:1CF mixture, suggesting a different heterotrimer composition (Fig. 7.8b). In addition, the ^{1}H,^{15}N–HSQC spectra of CK and the 2CK:1CF mixture demonstrated the appearance of new peaks in the 2CK:1CF mixture, supporting the successful formation of heterotrimers composed of CK and CF (Fig. 7.8c). Remarkably, both the 2CK:1CF and 1CK:2CF mixtures exhibited almost identical ^{1}H,^{15}N–HSQC spectra, indicating the formation of the same major heterotrimer $(CK)_2(CF)_1$ in both mixtures (Fig. 7.8d). NOESY and TOCSY spectra were recorded for the 2CK:1CF mixture, revealing NOE signals between the aromatic hydrogens of Phe and the side chain hydrogens of Lys, supporting the existence of cation –π interactions between Lys and Phe within the heterotrimer.

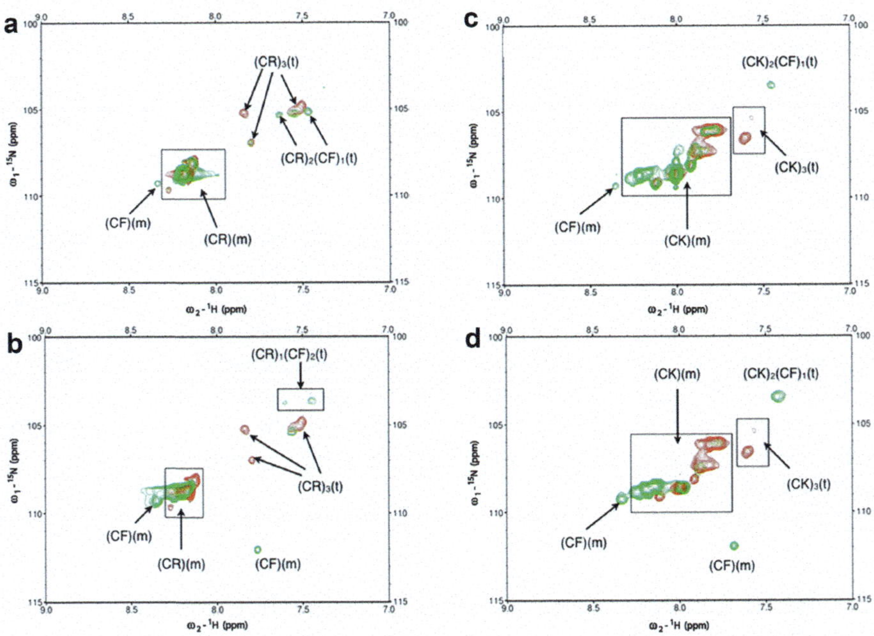

Fig. 7.8 Overlapped ^{1}H,^{15}N–HSQC of (**a**) CR (green) and 2CR:1CF (red) and (**b**) CR (green) and 1CR:2CF (red). Overlapped ^{1}H,^{15}N–HSQC of (**c**) CK (green) and 2CK:1CF (red) and (**d**) CK (green) and 1CK:2CF (red). The "t" denotes trimer and "m" denotes monomer [38]. Reprinted from ref. [38], Copyright © 2017, with permission from American Chemical Society

7.5.2 NMR Characterization of ABC-Type Heterotrimers

Hartgerink and co-workers designed a single-composition ABC heterotrimer using electrostatic interactions between oppositely charged amino acids, with peptides A (sequence YG(PKGPKGPKGDKGPKG)$_2$), B (sequence YG(DOG)$_{10}$), and C (sequence YG(POGPKGPOGPOGPOG)$_2$) to stabilize the desired species and destabilize other competing states [39]. The ^{1}H,^{15}N–HSQC spectrum of peptide C exhibited distinct peaks corresponding to the C·C·C homotrimer (G_H) with the expected chemical shift (Fig. 7.9a), while the ^{1}H,^{15}N–HSQC spectrum of the A, B, and C mixture showed three characteristic peaks for the triple-helical conformation, which were distinct from those of the C·C·C homotrimer (Fig. 7.9b). However, in addition to the peaks corresponding to the principal register of the ABC system, an additional cross-peak (G'_B) was detected in the spectrum, which was attributed to a competing register of the ABC heterotrimer (Fig. 7.9b).

Peptide B1 (sequence YG(EOG)$_{10}$) was synthesized to explore the influence of the B chain on heterotrimer formation, and the ^{1}H,^{15}N–HSQC spectrum of the A-B1-C mixture demonstrated the exclusive presence of the AB1C heterotrimer, indicating that peptide B1, containing Glu, exhibits enhanced specificity compared to peptide B with Asp (Fig. 7.9c). In the NOESY spectrum of the A-B1-C peptide mixture, a distinct cross peak at 8.61 and 3.02 ppm between the aspartate amide proton in chain A and the lysine ε-methylene in chain C suggested the formation of salt bridges (Fig. 7.9d). Furthermore, the peaks observed at 3.68 and 3.02 ppm were attributed to the lysine ε-protons in chain C and the glycine α-protons preceding the aspartate in chain A, respectively, providing compelling evidence that the triple helix register was A·B1·C (Fig. 7.9d). The rationally designed system for single-composition ABC heterotrimers sheds light on synthetic collagen triple helix self-assembly and collagen assembly mechanisms.

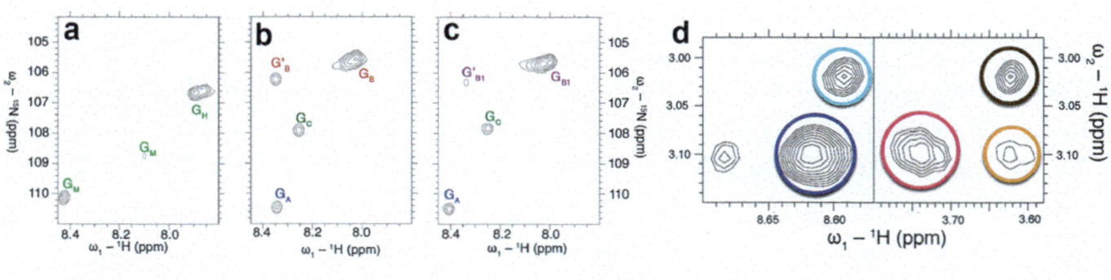

Fig. 7.9 ^{1}H,^{15}N-HSQC spectra of the (**a**) C·C·C, (**b**) A·B·C, and (**c**) A·B1·C triple helices at 25 °C. H indicates homotrimer while M indicates monomer of C·C·C. (**d**) NOESY spectrum of the A·B1·C heterotrimer highlighting interstrand interactions involving the lysine side chains and aspartate/glutamate and glycine backbone atoms [39]. Reprinted from ref. [39], Copyright © 2012, with permission from American Chemical Society

7.5.3 *NMR Characterization of AAB Heterotrimers Modeling Natural Interruptions*

Peptides A and B were designed to model natural interruptions in the $[\alpha_1(IV)]_2\alpha_2(IV)$ isoform of type IV collagen. Peptide A contained residues 289–299 from the human $\alpha_1(IV)$ chain with a G1G interruption (with sequence of Ac-OGPKGPOGPOGMPGVGEKGEOGEOGPOGPOGY-NH_2) and selectively $^{15}N/^{13}C$-labeled GVG at the interruption site, as well as labeled Gly25 at the C terminus [40]. Peptide B included residues 288–295 from the human $\alpha_2(IV)$ chain with a GISLKG interruption (sequence: Ac-POGDOGPOGPOGISLKGEEGPOGPAGPOGYOG-NH_2) and selectively $^{15}N/^{13}C$-labeled ISL at the interruption site, along with labeled Ala25 at the C terminus.

NMR studies were conducted on the 2A:1B peptide mixture, along with A homotrimers and B homotrimers, to assess the formation of heterotrimers. The HSQC spectrum of homotrimer peptide A (green) exhibited three trimer peaks for labeled residues Val15, Gly16, and Gly25, indicating distinct chemical environments in three different chains (Fig. 7.10a) [40]. The HSQC spectrum of homotrimer peptide B (blue) displayed three trimer peaks for Leu15, indicating distinct environments for the Leu residues located in the interruption region in each of the three chains of the homotrimer (Fig. 7.10b). In contrast, the HSQC spectrum of the mixture showed two trimer resonances (^{A}G14, ^{A}V15, ^{A}G16, and ^{A}G25) for all labeled residues in peptide A, while peptide B displayed a single trimer resonance (^{B}I13, ^{B}S14, ^{B}L15 and ^{B}A25), indicating a heterotrimer composition of two A chains and one B chain. The chemical shift differences in the trimer peaks between the homotrimers and the 2A:1B mixture highlighted distinct environments for residues in polypeptide chains A and B in homotrimers versus heterotrimers, exemplified by the trimer resonances of Gly25 (^{A}G25) and Ile13 (^{B}I13) in (Fig. 7.10c, d). The absence of any trimer resonance corresponding to the chemical shifts observed in the homotrimers for the labeled residues in the peptide mixture indicated the formation of a single-component heterotrimer (2A:1B) without any observable homotrimers.

Chain assignments for the 2A:1B peptide were based on interchain NOEs, summarized in the experimental contact map (Fig. 7.11a) [40]. The observed interchain NOEs, including ^{A}G14 NH–^{B}I13 αH, ^{A}V15 NH–^{B}I13 αH, ^{A}G16 NH–^{B}S14 αH, ^{A}G14 NH–^{A}V15 αH, and ^{B}I13 NH–^{A}P13 αH, unequivocally established that the A chain (violet) served as the leading chain, while the B chain acted as the lagging chain, confirming that the peptide 2A:1B mixture formed a single-component heterotrimer with the chain register AAB. Amide proton temperature gradients were measured for the homotrimers A_3 and B_3, as well as the heterotrimer AAB (Fig. 7.11b, d). The amide proton temperature gradients for all labeled residues at the interruption sites (Gly14-Val15-Gly16 in homotrimer A and Ile13-Ser14-Leu15 in homotrimer B) were more negative than -6, indicating the absence of hydrogen bonds (Fig. 7.11b, c). Leu15 in the B chain showed a temperature gradient of –2.8, indicating hydrogen

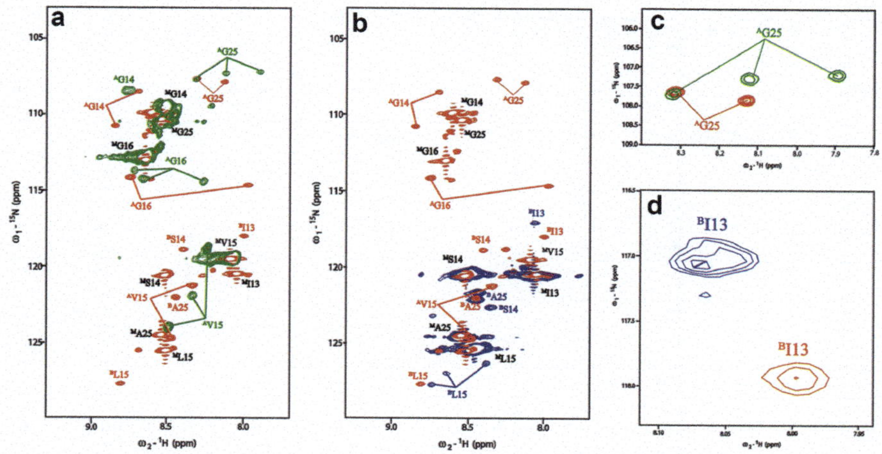

Fig. 7.10 **a** Overlapped ^{1}H–^{15}N HSQC spectra of AAB (red) and A (green). **b** Overlapped ^{1}H–^{15}N HSQC spectra of AAB (red) and B (blue). **c** Zoomed-in spectra of ^{A}G25 (colored red in AAB and green in A). **d** Zoomed-in spectra of ^{B}I13 (colored red in AAB and blue in B). Superscript A and B indicate the trimer resonances for peptide chains A and B, respectively, whereas superscript M represents the monomer resonances [40]. Reprinted from ref. [40], Copyright © 2015, with permission from Elsevier Science Ltd.

bonding, and Gly14 residues in the two A chains had gradients of –4.4 and –4.6, suggesting potential hydrogen bond formation (Fig. 7.11d). The flexibility of residues within the AAB heterotrimer was assessed using ^{1}H–^{15}N NOE values (Fig. 7.11e). Most labeled residues, except ^{A}G16, displayed NOE values larger than 0.6, indicating a rigid nature for most residues at the interruption sites, except for the two Gly16 residues at the C terminus of the GVG interruption sites in the A chains (Fig. 7.11e). These findings offer novel insights into the conformation and dynamics of the G1G/G4G interruptions within a heterotrimeric environment in type IV collagen.

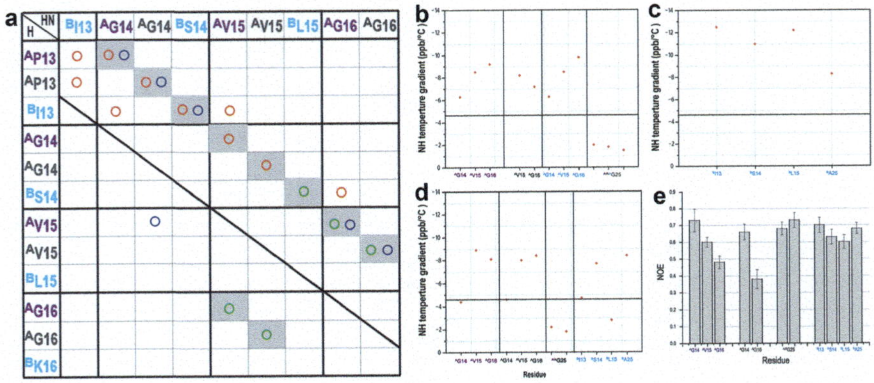

Fig. 7.11 **a** Contact map generated from NH–H experimental NOEs of peptide mixture 2A:1B. Experimental NOEs are represented by circles: HN–HN (green circle); HN–HA (red circle); and HN-side chain protons (blue circle). The shaded squares in the background denote intrachain NOEs. Amide proton temperature gradients of homotrimer A (**b**), homotrimer B (**c**), and heterotrimer AAB (**d**). The black horizontal line corresponds to a value of –4.6, the cut-off for hydrogen bonding, with less negative values indicative of hydrogen bonding. **e** Heteronuclear $^{1}H–^{15}N$ NOE values of heterotrimer peptide AAB. Violet and gray colors represent the leading and middle chain A, respectively, whereas the azure color represents the lagging chain B [40]. Reprinted from ref. [40], Copyright © 2015, with permission from Elsevier Science Ltd.

References

1. Hu Y, Cheng K, He L et al (2021) NMR-based methods for protein analysis. Anal Chem 93:1866–1879
2. Fernández C, Wüthrich K (2003) NMR solution structure determination of membrane proteins reconstituted in detergent micelles. FEBS Lett 555:144–150
3. Huang C, Kalodimos CG (2017) Structures of large protein complexes determined by nuclear magnetic resonance spectroscopy. Annu Rev Biophys 46:317–336
4. Jarymowycz VA, Stone MJ (2006) Fast time scale dynamics of protein backbones: NMR relaxation methods, applications, and functional consequences. Chem Rev 106(5):1624–1671
5. Lakomek NA, Ying J, Bax A (2012) Measurement of 15 N relaxation rates in perdeuterated proteins by TROSY-based methods. J Biomol NMR 53:209–221
6. Sapienza PJ, Lee AL (2010) Using NMR to study fast dynamics in proteins: methods and applications. Curr Opin Pharmacol 10(6):723–730
7. Walters BT, Ricciuti A, Mayne L et al (2012) Minimizing back exchange in the hydrogen exchange-mass spectrometry experiment. J Am Soc Mass Spectrom 23(12):2132–2139
8. Li R, Woodward C (1999) The hydrogen exchange core and protein folding. Protein Sci 8(8):1571–1590
9. Kateb F, Pelupessy P, Bodenhausen G (2007) Measuring fast hydrogen exchange rates by NMR spectroscopy. J Magn Reson 184(1):108–113
10. Marion D (2013) An introduction to biological NMR spectroscopy. Mol Cell Proteomics 12(11):3006–3025
11. Coen M, Holmes E, Lindon JC et al (2008) NMR-based metabolic profiling and metabonomic approaches to problems in molecular toxicology. Chem Res Toxicol 21(1):9–27
12. Zhuravleva A, Korzhnev DM (2017) Protein folding by NMR. Prog Nucl Magn Reson Spectrosc 100:52–77
13. Clore GM, Gronenborn AM (1998) NMR structure determination of proteins and protein complexes larger than 20 kDa. Curr Opin Chem Biol 2:564–570

14. Güntert P (2009) Automated structure determination from NMR spectra. Eur Biophys J 38:129–143
15. Bothe JR, Nikolova EN, Eichhorn CD et al (2011) Characterizing RNA dynamics at atomic resolution using solution-state NMR spectroscopy. Nat Methods 8(11):919–931
16. Gaponenko V, Howarth JW, Columbus L et al (2000) Protein global fold determination using site-directed spin and isotope labeling. Protein Sci 9(2):302–309
17. Tang C, Gong Z (2020) Integrating non-NMR distance restraints to augment NMR depiction of protein structure and dynamics. J Mol Biol 432(9):2913–2929
18. Prestegard JH, Bougault CM, Kishore AI (2004) Residual dipolar couplings in structure determination of biomolecules. Chem Rev 104(8):3519–3540
19. Clore GM, Tang C, Iwahara J (2007) Elucidating transient macromolecular interactions using paramagnetic relaxation enhancement. Curr Opin Struct Biol 17(5):603–616
20. Momot KI (2022) Hydrated collagen: where physical chemistry, medical imaging, and bioengineering Meet. J Phys Chem B 126(49):10305–10316
21. Fields GB (2010) Synthesis and biological applications of collagen-model triple-helical peptides. Org Biomol Chem 8(6):1237–1258
22. Berisio R, De Simone A, Ruggiero A et al (2009) Role of side chains in collagen triple helix stabilization and partner recognition. J Peptide Sci Official Publ Euro Peptide Soc 15(3):131–140
23. Werner JM, Knott V, Handford PA et al (2000) Backbone dynamics of a cbEGF domain pair in the presence of calcium. J Mol Biol 296(4):1065–1078
24. Millhauser GL (1995) Views of helical peptides: a proposal for the position of 310-helix along the thermodynamic folding pathway. Biochemistry 34(12):3873–3877
25. Salvador P (2014) Dependencies of J-couplings upon dihedral angles on proteins. In: Annual reports on NMR spectroscopy, vol 81. Academic, pp 185–227
26. Bodian DL, Madhan B, Brodsky B et al (2008) Predicting the clinical lethality of osteogenesis imperfecta from collagen glycine mutations. Biochemistry 47:5424–5432
27. Yu YC, Roontga V, Daragan VA et al (1999) Structure and dynamics of peptide− amphiphiles incorporating triple-helical proteinlike molecular architecture. Biochemistry 38(5):1659–1668
28. Kramer RZ, Bella J, Mayville P et al (1999) Sequence dependent conformational variations of collagen triple-helical structure. Nat Struct Mol Biol 6:454–457
29. Engel J, Bächinger HP (2005) Structure, stability and folding of the collagen triple helix. In: Collagen: primer in structure, processing and assembly, vol 247, pp 7–33
30. Long CG, Braswell E, Zhu D et al (1993) Characterization of collagen-like peptides containing interruptions in the repeating Gly-XY sequence. Biochemistry 32(43):11688–11695
31. Bella J, Liu J, Kramer R et al (2006) Conformational effects of Gly-X-Gly interruptions in the collagen triple helix. J Mol Biol 362(2):298–311
32. Li Y, Brodsky B, Baum J (2007) NMR shows hydrophobic interactions replace glycine packing in the triple helix at a natural break in the (Gly-X-Y) repeat. J Biol Chem 282:22699–22706
33. Wang AC, Bax A (1996) Determination of the backbone dihedral angles φ in human ubiquitin from reparametrized empirical karplus equations. J Am Chem Soc 118:2483–2494
34. Kramer RZ, Venugopal MG, Bella J et al (2000) Staggered molecular packing in crystals of a collagen-like peptide with a single charged pair. J Mol Biol 301:1191–1205
35. Lovell SC, Davis IW, Arendall III WB et al (2003) Structure validation by Cα geometry: ϕ,ψ and Cβ deviation. Proteins: Struct Funct Bioinf 50:437–450
36. Xiao J, Addabbo RM, Lauer JL et al (2010) Local conformation and dynamics of isoleucine in the collagenase cleavage site provide a recognition signal for matrix metalloproteinases*. J Biol Chem 285:34181–34190
37. Xiao J, Madhan B, Li Y et al (2011) Osteogenesis imperfecta model peptides: incorporation of residues replacing Gly within a triple helix achieved by renucleation and local flexibility. Biophys J 101:449–458
38. Chiang C-H, Fu Y-H, Horng J-C (2017) Formation of AAB-type collagen heterotrimers from designed cationic and aromatic collagen-mimetic peptides: evaluation of the C-terminal Cation−π interactions. Biomacromol 18:985–993

39. Fallas JA, Lee MA, Jalan AA, Hartgerink JD (2012) Rational design of single-composition ABC collagen heterotrimers. J Am Chem Soc 134:1430–1433
40. Xiao J, Sun X, Madhan B et al (2015) NMR studies demonstrate a unique AAB composition and chain register for a heterotrimeric type IV collagen model peptide containing a natural interruption site. J Biol Chem 290:24201–24209

Chapter 8
Fluorescence Detection of Collagen Peptides

Abstract Collagen peptides serve as crucial biomarkers in diverse pathologies, prompting the need for robust detection assays. Fluorescence, a non-destructive and highly sensitive optical technique reliant on fluorophore emission, enables precise biomolecule quantification. This chapter offers a comprehensive overview of customized fluorescence assays for collagen peptides, notably focusing on fluorescence self-quenching assays. These assays detect changes in fluorescence intensity resulting from close fluorophore proximity, facilitating the assessment of collagen peptide helix composition. Additionally, advanced methodologies employing short probe peptides and the detection of collagen peptides with single amino acid mutations are discussed. Furthermore, the chapter explores graphene oxide-based fluorescence quenching assays, including advancements like the GO paper-based FRET assay and a graphene oxide-aided triple helical aggregation-induced emission biosensor, alongside investigations into fluorescence assays utilizing novel 2D nanomaterials such as WS_2 and MoS_2. Detailed protocols for conducting both fluorescence self-quenching assays and GO-based fluorescence assays are provided.

Keywords Fluorescence assays · Self-quenching · Collagen peptides · Graphene oxide · 2D nanomaterials

8.1 Introduction to Fluorescence Assays of Collagen Peptides

Fluorescence is a non-destructive and highly sensitive technology that utilizes the emission of light from fluorophores upon excitation, enabling precise and sensitive detection and quantification of specific biomolecules associated with critical diseases [1–3]. It offers real-time data acquisition, allowing dynamic monitoring of biomarker levels in biological samples [4]. Moreover, fluorescence technology provides a wide range of fluorescent probes and dyes, enabling multiplexing to detect multiple biomarkers simultaneously [5]. With their versatility and high-throughput capability, fluorescence assays are invaluable tools for detecting various biomarkers, including

© Springer Nature Singapore Pte Ltd. 2024

J. Xiao, *Collagen Mimetic Peptides and Their Biophysical Characterization*, Springer Series in Biomaterials Science and Engineering 20,
https://doi.org/10.1007/978-981-10-1941-8_8

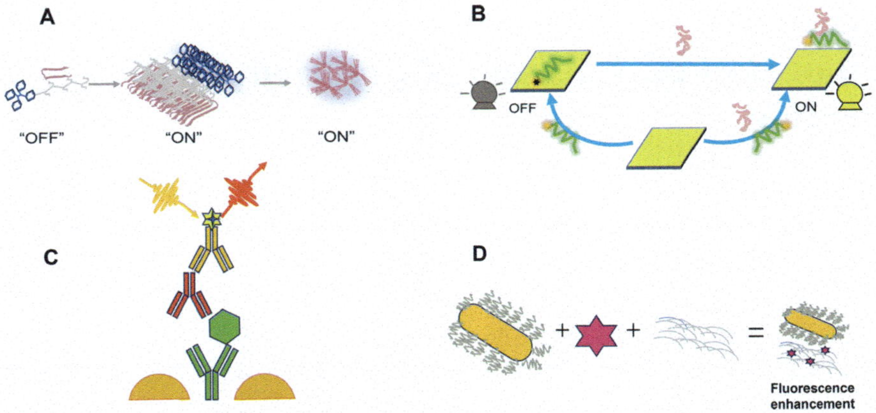

Fig. 8.1 Schematic diagrams of fluorescence-sensing assays

proteins and peptides, and hold promising applications in clinical diagnostics and disease monitoring (Fig. 8.1) [6–8].

Collagen peptides have gained widespread recognition as critical biomarkers in various debilitating diseases such as arthritis, fibrosis, and tumors [9–11]. Dysregulation of collagen metabolism leads to the excessive production of collagen fragments, releasing collagen peptides into the bloodstream and other biological fluids, contributing to a broad range of connective tissue disorders [12]. Notably, specific collagen fragments, such as C1M and C3M derived from type I and III collagen by MMP-2 and MMP-9, have been closely linked to liver fibrosis severity, while C2M, a fragment of type II collagen, has been established as a sensitive marker for cartilage erosion in arthritis [13–15]. Recent findings have revealed a high level of serum C1M to be associated with increased mortality in postmenopausal women, as observed in a large prospective cohort study [16]. The specific detection of these collagen-derived fragments is of utmost importance for precise diagnostics and prognostics, significantly improving the management and treatment strategies for various critical diseases.

Extensive efforts have been dedicated to developing efficient antibodies and peptides for detecting collagen fragments. Various approaches, such as using the humanized monoclonal IgG1 antibody (D93) to target denatured collagen type IV by binding to a cryptic site of MMP-processed collagen, and employing monoclonal antibodies E1E5 and E4A11 for specific binding to denatured collagen type II through ELISA-based screening, have been explored [17, 18]. Additionally, the peptide sequence TLTYTWS has been identified to interact specifically with a cryptic site on MMP-2 processed type IV collagen, as revealed by phage display library screening [19]. However, these methods have encountered challenges due to their time-consuming procedures, low specificity, poor binding affinity, and the technical difficulties of discovering novel antibodies for collagen because of its repetitive

$(Gly\text{-}X\text{-}Y)_n$ amino acid sequences. Hence, the development of robust assays for the sensitive and specific detection of collagen peptides remains an urgent goal.

8.2 Fluorescence Self-quenching Assay of a Classic Collagen Peptide

8.2.1 Introduction to Fluorescence Self-quenching Assay

Fluorescence quenching assay is a robust technique that exploits the phenomenon of fluorescence quenching, where the emission of light from a fluorophore is reduced upon interaction with a quencher molecule [20, 21]. This assay often relies on Förster resonance energy transfer (FRET), a non-radiative energy transfer process that occurs when two fluorophores are in close proximity. During FRET, the energy from an excited donor fluorophore is transferred to an acceptor fluorophore, resulting in emission from the acceptor fluorophore instead of the donor fluorophore [22]. The fluorescence quenching assay, utilizing FRET, empowers researchers to explore diverse molecular interactions, such as ligand-receptor binding, protein–protein interactions, and enzymatic activity, while quantitatively analyzing their binding affinities and kinetics through the measurement of decreased fluorescence intensity [23–25].

Fluorescence self-quenching assay, also referred to as concentration-dependent fluorescence quenching, is a fluorescence-based technique that detects changes in fluorescence intensity caused by the close proximity of multiple fluorophores within a confined space [26]. This phenomenon occurs when the fluorophores become more densely packed, leading to a reduction in fluorescence emission. The fluorescence self-quenching assay simplifies the labeling procedure compared to FRET, as it requires conjugation of only one type of fluorophore instead of two. Fluorescence self-quenching has emerged as a powerful tool in the study of single protein systems, providing valuable insights into protein folding, conformational fluctuations of unfolded proteins, and amyloid β (Aβ) peptide aggregation [27].

Our research group has recently developed a fluorescence self-quenching assay to determine the helix composition of collagen peptides [28]. In this assay, FAM, a convenient fluorescein-based fluorescent dye, is covalently attached to a collagen mimetic peptide. It is hypothesized that the FAM-labeled peptide exhibits strong fluorescence in the monomer state, while the formation of homotrimers leads to the close proximity of the dye, resulting in fluorescence self-quenching. In contrast, when the FAM-labeled peptide forms heterotrimers with a complementary unlabeled collagen-like peptide, the strong fluorescence remains. The difference in fluorescence intensity allows the distinction between homotrimers and heterotrimers, facilitating the accurate determination of the helix composition of collagen-like peptides (Fig. 8.2A).

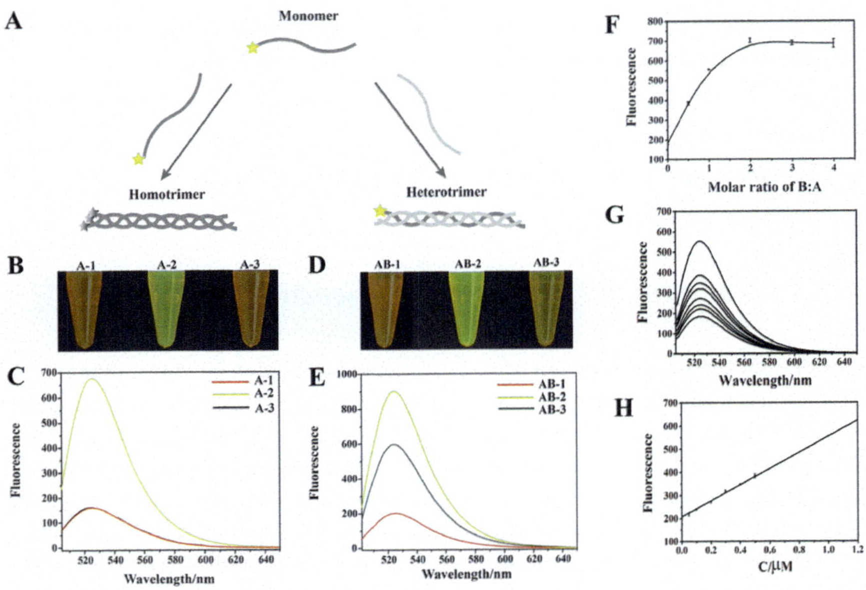

Fig. 8.2 **A** Schematic illustration of the fluorescence self-quenching assay to detect the helix composition of collagen mimetic peptides. Colorimetric (**B**) and fluorometric characterization (**C**) of peptide FAM-G(PRGPOG)$_5$. Colorimetric (**D**) and fluorometric characterization (**E**) of the mixture of peptide FAM-G(PRGPOG)$_5$ (denoted as peptide A) and G(POG)$_{10}$ (denoted as peptide B) at a molar ratio of 1: 2. **F** Fluorescence characterization of peptides A and B at different molar ratios (B: A = 0, 0.5, 1, 2, 3, 4).The fluorescence intensity (**G**) and linear fluorescence restoration (**H**) of FAM-labeled peptide A by hybridization with peptide B [28]. Reprinted from ref. [28], Copyright © 2016, with permission from Elsevier Science Ltd.

8.2.2 Fluorescence Self-quenching Assay for Monitoring the Triple Helix Composition

In a proof-of-concept application, two peptides, FAM-G(PRGPOG)$_5$ (peptide A) with FAM labeling and G(POG)$_{10}$ (peptide B) without labeling, were developed to explore the distinction in fluorescence self-quenching between homotrimeric and heterotrimeric helix formations (Fig. 8.2) [28]. Peptide A displayed distinct color and fluorescence intensity in its monomeric and homotrimeric states. The homotrimeric peptide A solution exhibited a peru color at 4 °C, while upon heating at 70 °C, it transformed into a yellow-colored monomer. Upon re-equilibration at 4 °C, peptide A reformed the homotrimeric state, resulting in the color reverting to peru (Fig. 8.2B). Additionally, the fluorescence intensity of the homotrimeric peptide A was significantly lower than that of the monomer, which was attributed to fluorescence self-quenching (Fig. 8.2C). These results illustrate the efficacy of colorimetric and fluorometric assays in identifying distinct states of collagen peptides.

Colorimetric and fluorometric assays were employed to investigate the formation of heterotrimer using peptide mixtures 1A:2B. Similar to peptide A alone, the

1A:2B mixture exhibited peru color and weak fluorescence at 4 °C, transitioning to yellow color and strong fluorescence upon heating at 70 °C. Upon re-equilibration at 4 °C, the solution displayed light yellow color and largely preserved fluorescence, suggesting the formation of heterotrimer rather than homotrimer peptide A (Fig. 8.2D, E). The fluorescence spectra of the A:B mixture at different molar ratios reached a plateau above 2, indicating the exact helix composition of the heterotrimer as ABB (Fig. 8.2F).

The capability of the fluorescence self-quenching assay to detect heterotrimers was investigated by hybridizing FAM-labeled peptide A with various concentrations of the target peptide B. As the concentration of the target peptide increased, the percentage of heterotrimer also increased, leading to a corresponding increase in fluorescence intensity. The fluorescence intensity exhibited a linear relationship with the concentration of the target peptide ($R^2 = 0.99$) within a broad linear range from 50 to 1000 nM (Fig. 8.2G, H). The study demonstrates the remarkable sensitivity of the fluorescence self-quenching assay in detecting heterotrimer formation at the ultra-low concentration of nM, significantly surpassing the capabilities of current techniques for characterizing collagen triple helices. By utilizing a dye-labeled collagen-like probe peptide, this novel approach opens new avenues for screening potential target sequences suitable for constructing heterotrimers with the probe peptide, addressing the longstanding challenge of heterotrimer model design for collagen. The fluorescence self-quenching assay proves to be a versatile tool applicable to various hybrid systems, facilitating an unequivocal determination of the chain composition within a heterotrimer triple-helix.

8.3 Advanced Fluorescence Self-quenching Assay of Collagen Peptides

8.3.1 Fluorescence Self-quenching Assay Using a Short Probe Peptide

Our research group extensively investigated the targeting efficacy of a series of probe peptides, FAM-(Pro-Hyp-Gly)$_n$ (n = 4, 6, 8, 10), by utilizing the fluorescence self-quenching assay [29]. To screen and identify an optimal probe for detecting collagen sequences, target sequences with different lengths of (Pro-Hyp-Gly)$_n$ (n = 4, 6, 8, 10) without any fluorescent labels were devised. In the self-quenching assay, the FAM-labeled probe peptide exhibits robust fluorescence upon forming a heterotrimer with the unlabeled target peptide, whereas it self-assembles into a homotrimer, leading to fluorescence self-quenching when it does not match the target peptide (Fig. 8.3A). The variation in fluorescence intensity of the peptide mixture therefore enables the identification of the precise probe peptide that effectively binds to the target peptide.

The three peptide probes, FAM-G(POG)$_6$, FAM-G(POG)$_8$, and FAM-G(POG)$_{10}$, all exhibited significantly reduced fluorescence intensity at lower temperatures,

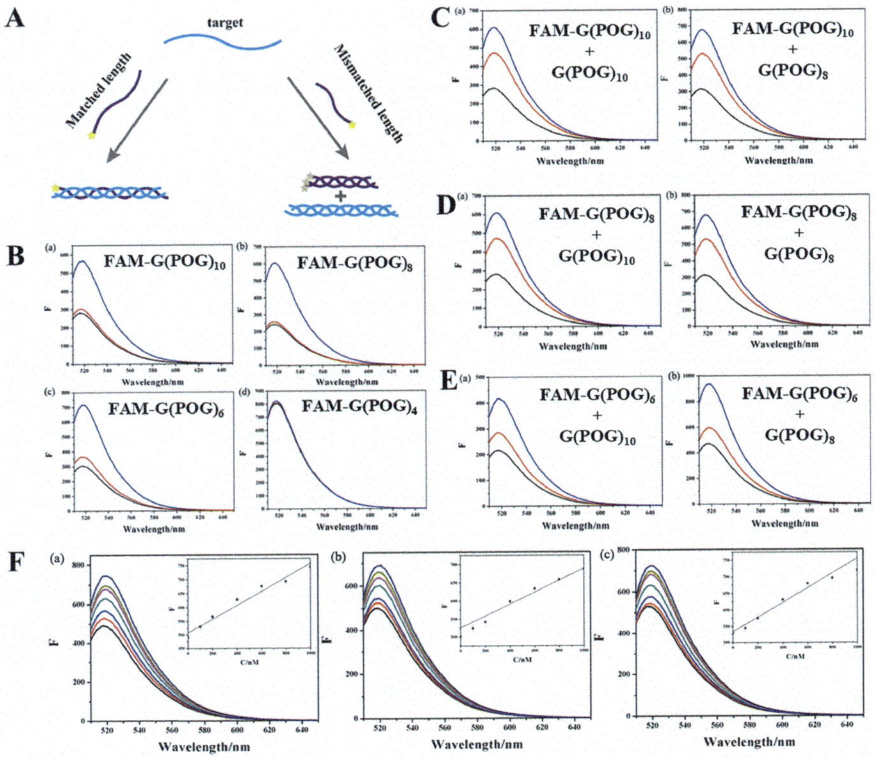

Fig. 8.3 **A** Schematic illustration of monitoring the interaction of target peptides and fluorescent probe peptides of different lengths using fluorescence self-quenching assay. **B** Fluorescence characterization of FAM-G(POG)$_{10}$, FAM-G(POG)$_8$, FAM-G(POG)$_6$, and FAM-G(POG)$_4$. **C** Fluorescence characterization of the mixture of probe peptide FAM-G(POG)$_{10}$ with various target peptides G(POG)$_{10}$ and G(POG)$_8$. **D** Fluorescence characterization of the mixture of probe peptide FAM-G(POG)$_8$ with various target peptides G(POG)$_{10}$ and G(POG)$_8$. **E** Fluorescence characterization of the mixture of probe peptide FAM-G(POG)$_6$ with various target peptides G(POG)$_{10}$ and G(POG)$_8$. Fluorescence profiles are recorded for the peptide solutions incubated at 4 °C (black), heated to 90 °C for 25 min (blue), and re-equilibrated at 4 °C (red). **F** The proportional fluorescence restoration of probe peptide FAM-G(POG)$_8$ by the hybridization with target collagen peptide sequence G(POG)$_{10}$ in PBS (**a**), human urine (**b**) and saliva (**c**) samples [29]. Reprinted from ref. [29], Copyright © 2018, with permission from Elsevier Science Ltd.

confirming their propensity for homotrimer formation, while the probe FAM-G(POG)$_4$ maintained elevated fluorescence intensity at low temperatures, indicating a limited tendency for homotrimer formation and unsuitability as a viable probe (Fig. 8.3B). Upon hybridization with the target sequences G(POG)$_{10}$ and G(POG)$_8$, both FAM-G(POG)$_{10}$ and FAM-G(POG)$_8$ peptide probes exhibited a significant increase in fluorescence intensity as compared to their individual states at low temperature, underscoring their proficiency in recognizing and binding to the target collagen sequences, while the FAM-G$(POG)_6$ probe displayed comparatively low fluorescence both before and after hybridization with target sequences, suggesting its limited ability to bind with the target collagen sequences and form heterotrimers (Fig. 8.3C–E). These findings unveiled that among the four peptide probes screened, FAM-G$(POG)_8$ emerges as the shortest yet remarkably proficient candidate for effectively detecting target collagen sequences.

The fluorescence probe FAM-G$(POG)_8$'s capacity to detect the target collagen sequence G$(POG)_{10}$ was further examined, demonstrating a linear relationship ($R^2 = 0.99$) between fluorescence intensity and target peptide concentration across an expansive range from 100 to 1000 nM, facilitating accurate detection down to 84 nM, similar to the FAM-G$(POG)_{10}$-based fluorescence assay (Fig. 8.3F). The fluorescence self-quenching assay, combined with the FAM-G$(POG)_8$ probe peptide, was extended to the detection of target collagen sequences in human urine and saliva samples, resulting in a similar detection limit and a parallel linear correlation across the 100–1000 nM range, effectively showcasing the capability of the FAM-G$(POG)_8$ probe to identify target collagen peptides within complex biological matrices (Fig. 8.3F). The fluorescence self-quenching assay presents a potent and innovative approach for the exploration of efficient peptides in recognizing collagen biomarkers, with the prospect of extending its utility to identify probe peptides targeting various protein biomarkers associated with pathological states.

8.3.2 Detection of Target Collagen Peptides with Single Amino Acid Mutation

Our research group has recently established a fluorescence self-quenching assay employing the peptide probe FAM-G$(POG)_{10}$ to selectively target collagen peptide sequences bearing OI mutations [30]. A series of host–guest peptides was crafted to emulate pathological Gly-X mutations within a precisely defined amino acid sequence context, where the host peptide G (with amino acid sequence G$(POG)_4$POG$(POG)_5$) mirrored prevalent POG sequences in native collagen, while peptides A (with amino acid sequence G$(POG)_4$POA$(POG)_5$), S (with amino acid sequence G$(POG)_4$POS$(POG)_5$), and R (with amino acid sequence G$(POG)_4$POR$(POG)_5$) incorporated Gly-Ala, Gly-Ser, and Gly-Arg mutations, respectively. Hybridizing the probe peptide with the host peptide in a natural collagen sequence is anticipated to lead to regular heterotrimer formation, thereby maintaining robust fluorescence; however, the probe peptide's capability to engage in heterotrimer formation with the mutation peptide is considerably hindered, promoting the emergence of homotrimer and heterotrimer structures, ultimately resulting in a noteworthy decrease in fluorescence intensity due to the self-quenching of the FAM dye within a homotrimeric environment (Fig. 8.4A).

Upon hybridization with the peptide probe FAM-G$(POG)_{10}$, the mutant peptide A displayed a lower fluorescence intensity than the host peptide G, indicating a higher

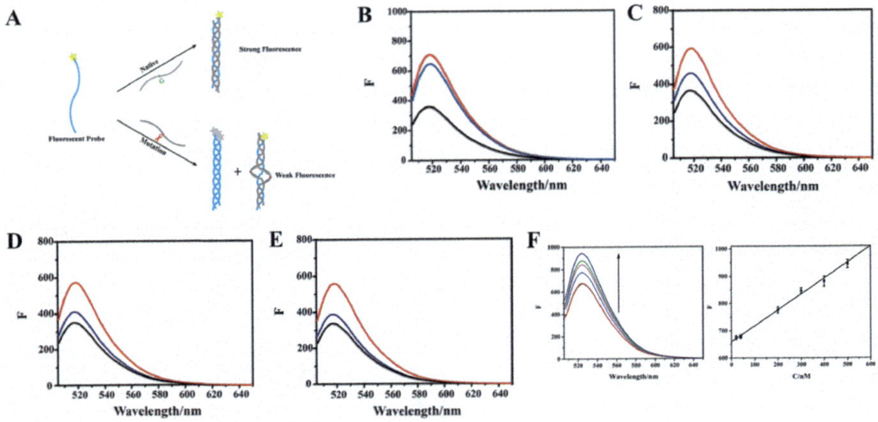

Fig. 8.4 **A** Schematic illustration of the fluorescence self-quenching assay to detect target collagen peptides with a single Gly–X mutation. Fluorescence profiles of the 1:2 mixtures of probe peptide P1 hybridized with peptide G (**B**), peptide A (**C**), peptide S (**D**), peptide R (**E**). The emission spectra were recorded at an excitation wavelength of 497 nm for each mixture under three conditions including incubation at 4 °C (black), heating to 90 °C for 20 min (red), and re-equilibration at 4 °C (blue). **F** The proportional fluorescence restoration of the probe peptide P1 by the hybridization with the mutation peptide A in PBS buffer [30]. Reprinted from ref. [30], Copyright © 2019, with permission from Royal Society of Chemistry

prevalence of homotrimers and a reduced occurrence of heterotrimers, thus underscoring the diminished binding capacity of peptide A in comparison to the natural peptide G (Fig. 8.4B, C). Furthermore, peptides S and R exhibited significantly more substantial fluorescence reduction than peptide A, indicating an even weaker binding capacity for peptides S and R (Fig. 8.4D, E). The observed outcomes established a distinct hierarchy of binding capability of the probe peptide with respect to the target peptides, ranking as G > A > S–R, suggesting that the extent of fluorescence quenching was contingent upon the specific residue replacing Gly.

The discernible variance in fluorescence intensity allowed for the discrimination between mutation peptides and the native peptide, and the assay was employed for the quantitative detection of mutant peptide A. The augmentation in peptide A concentration induced heightened heterotrimer formation, as reflected in an increase in fluorescence intensity, underscoring a robust linear relationship between measured fluorescence intensity and peptide A concentration (Fig. 8.4F). Encompassing a linear range from 20 to 500 nM and showcasing precise detection of peptide A at concentrations as low as 17 nM, this fluorescence assay robustly demonstrates its aptitude in identifying collagen peptides harboring a solitary Gly mutation at the nanomolar scale (Fig. 8.4F). The fluorescence self-quenching assay significantly bolsters our ability to sensitively and accurately pinpoint collagen fragments featuring a single amino acid mutation, potentially accelerating the advancement of effective diagnostic approaches for collagen-related disorders.

8.4 Graphene Oxide-Based Fluorescence Assays for Collagen Peptides

8.4.1 Introduction to Graphene Oxide-Based Fluorescence Quenching Assays

Graphene oxide (GO), a two-dimensional carbon material consisting of a single layer of carbon atoms arranged in a honeycomb lattice, has attracted significant attention in the scientific community due to its remarkable mechanical, electrical, and optical properties [31]. With its atomically thin structure, large specific surface area, and distinctive electronic configuration, GO is well-equipped to interact effectively with both light and molecules, making it an ideal candidate for fluorescence-based sensing applications. Fluorescence quenching, which involves the suppression of fluorescence emission upon interaction with a quencher molecule, finds a powerful companion in GO thanks to its remarkable quenching efficacy attributed to robust π–π stacking interactions and energy transfer capabilities [32]. Additionally, the aromatic and electrically charged nature of GO enhances its potential by enabling efficient absorption of biomolecules through π–π or electrostatic interactions [33]. The integration of GO into fluorescence quenching assays holds significant promise for advancing biomolecule sensing platforms.

Leveraging the distinct properties of GO and incorporating dye-labeled single-stranded DNA (ssDNA), a variety of fluorescent biosensors have been engineered for DNA detection [34]. This approach capitalizes on the strong adsorption of ssDNA onto the GO surface, in contrast to the restricted interaction observed with double-stranded DNA (dsDNA) due to the encapsulation of exposed ssDNA bases within the helical structure. The hybridization of dye-labeled ssDNA with its complementary DNA sequence has demonstrated itself as a versatile and efficient technique for the precise and selective detection of various DNA targets. Although the widely recognized double helix structure has been extensively utilized to develop highly specific GO-based biosensors for DNA detection, the distinctive triple helix structure remains largely unexplored as a potential framework to construct similar sensing platforms for collagen.

Our research group has recently introduced GO-based fluorescence quenching assays, offering a highly sensitive and selective means for detecting collagen peptides (Fig. 8.5A) [35]. In this assay, a dye-labeled positively charged single-stranded collagen sequence (ssCOL) is designed as a probe peptide for the detection of collagen target peptides. The underlying hypothesis posits that the dye-labeled ssCOL exhibits robust fluorescence, and upon interaction, the negatively charged GO readily binds to the positively charged ssCOL, thereby leading to the quenching of the dye's fluorescence. In contrast, the presence of a single-stranded target collagen peptide triggers the formation of a heterotrimer with the dye-labeled ssCOL probe,

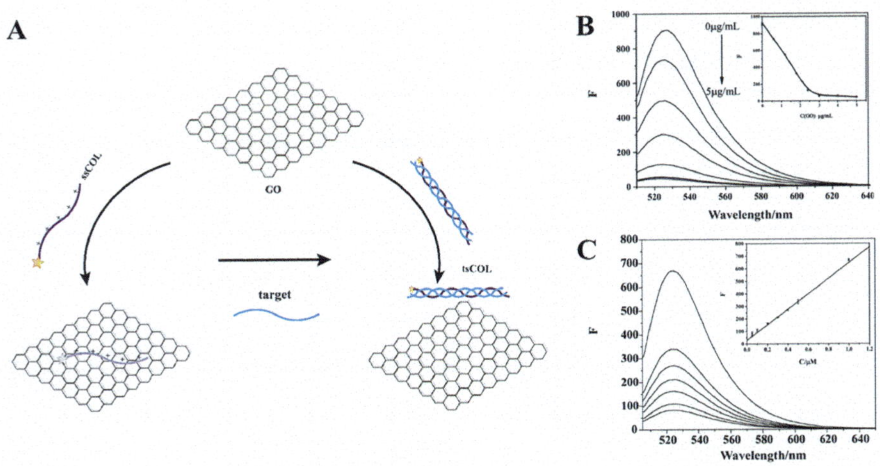

Fig. 8.5 **A** Schematic iillustration of a novel graphene oxide-based fluorescent assay utilizing the hybridization of a highly positively charged peptide with a complementary target collagen sequence. **B** The fluorescence quenching of the peptide probe FAM–GPR by GO. **C** The linear fluorescence restoration of the probe FAM–GPR by the hybridization with target peptide GPO in PBS buffer [35]. Reprinted from ref. [35], Copyright © 2016, with permission from Royal Society of Chemistry

causing the detachment of the probe from the GO surface and a consequential significant recovery of fluorescence (Fig. 8.5A). The discernible disparity in fluorescence intensity enables precise quantification of collagen target peptides.

8.4.2 *GO-Based Platform for Sensing Collagen Triple Helix*

In a proof-of-concept application, peptide FAM-GPR (FAM-G(PRGPOG)$_5$) was synthesized to detect an unlabeled collagen peptide GPO (G(POG)$_{10}$), through the assistance of GO (Fig. 8.5A) [35]. In the absence of GO, the single-stranded collagen (ssCOL) peptide probe FAM-GPR exhibited strong fluorescence at 524 nm, which gradually diminished as GO was progressively added, ultimately resulting in near-complete quenching of FAM-GPR fluorescence at a GO concentration of 3 mg/mL (Fig. 8.5B). The pronounced quenching efficiency of GO demonstrated the effective design of the ssCOL fluorescent peptide probe, showcasing its robust adsorption onto the GO surface.

To evaluate the quantitative potential of the GO-based platform for detecting collagen-like triple helical molecules, the ssCOL peptide probe FAM–GPR was subjected to hybridization with varying concentrations of the target collagen peptide GPO (Fig. 8.5B). Remarkably, the hybridization of FAM-GPR with peptide GPO led to a significant fluorescence recovery, underscoring the formation of a triple helix structure resulting from the interaction of FAM-GPR with the unlabeled peptide GPO, which exhibited notably weaker binding to GO. Higher GPO concentrations led to increased hybridization with FAM–GPR, intensifying retained fluorescence and showing a strongly proportional relationship between GPO concentration and fluorescence intensity. The ssCOL sensing platform exhibited a broad linear range of 50–1000 nM, achieving accurate detection of the target GPO at concentrations as low as 44 nM (Fig. 8.5C). The GO-based fluorescence assay exhibits superior sensitivity and selectivity to the target collagen peptides, with minimal interference from other proteins. Its versatility could extend to screening multiple target collagen sequences for new biomarkers and holds significant promise in molecular diagnostics of fibroproliferative diseases.

8.5 2D Nanomaterials-Based Fluorescence Assay of Collagen Peptides

8.5.1 GO Paper-Based FRET Assay for Collagen Peptides

The adoption of paper as a versatile substrate for biosensor construction has gained increasing attention, driven by its large surface area, cost-effectiveness, and easy fabrication [37]. Concurrently, the exploration of smartphones and digital cameras as detection tools in diagnostic assays has surged, capitalizing on their high resolution, micro-miniaturization, and affordability. Despite the substantial enhancement in analytical capabilities offered by advanced detection techniques like microarrays and next-generation sequencing, their widespread use has been impeded by the high cost of instrumentation and limited accessibility. The rise of paper- and smartphone-based devices has the potential to address the growing need for diagnostics, not only in routine situations but also in resource-constrained private and public environments.

Our research group has innovated a paper-based fluorescence resonance energy transfer (FRET) assay for rapid, cost-effective, and direct collagen triple helix detection (Fig. 8.6A) [37]. In this method, the dye-labeled probe peptide FAM–GPR (with sequence FAM–G(PRGPOG)$_5$) is designed to bind onto the GO-immobilized paper, leading to the quenching of dye fluorescence. In contrast, the hybridization of the probe peptide with the target collagen peptide GPO (with sequence G(POG)$_{10}$) induces the detachment of the probe peptide from GO, consequently restoring fluorescence (Fig. 8.6A). This strategy combines a GO-based fluorescence assay with smartphone imaging, creating a user-friendly and portable device that enables a convenient quantitative assay for collagen target peptides.

Fluorescence intensities of the probe peptide FAM-GPR were measured on paper strips modified with different concentrations of GO (0–0.1 mg/mL) (Fig. 8.6B). The fluorescence intensity of FAM-GPR on the paper gradually decreased as the amount of GO used for strip modification increased, with a significant quenching of the peptide's fluorescence observed at a GO concentration of 0.1 mg/mL (Fig. 8.6B). By subjecting the probe FAM-GPR to hybridization with varying concentrations of the

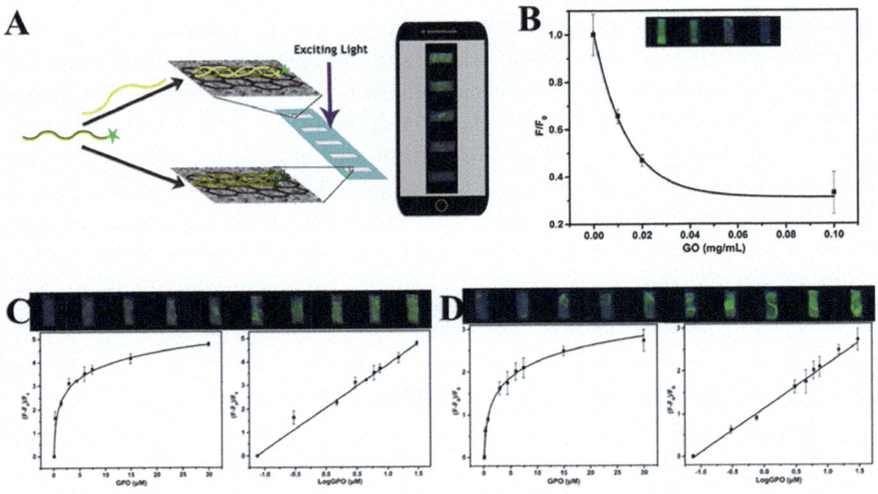

Fig. 8.6 **A** Schematic illustration of GO-based paper device for the detection of collagen peptide. **B** The fluorescence quenching of the probe peptide FAM-GPR by GO-immobilized paper. The relationship between the fluorescence intensity and the concentration of the target collagen peptide in phosphate buffer (**C**) and saliva (**D**) [37]. Reprinted from ref. [37], Copyright © 2016, with permission from Royal Society of Chemistry

target peptide GPO, a direct relationship between peptide concentration and fluorescence recovery of FAM-GPR on the paper device was established, showcasing the potential of the GO-immobilized paper platform for quantitative detection of target collagen peptides across a broad range from 75 nM to 30 mM (Fig. 8.6C). Furthermore, quantitative assays conducted on human saliva samples demonstrated the applicability of the GO-immobilized paper device for examining the target collagen peptide across a wide concentration range (75 nM–30 mM) within complex biological fluids (Fig. 8.6D). The amalgamation of the GO-based FRET assay with patterned paper introduces a potent tool for collagen molecule detection, offering several advantages such as minimal sample volumes, a multichannel detection mode, easy operation, and cost-effective equipment.

8.5.2 A Graphene Oxide-Aided Triple Helical Aggregation-Induced Emission Biosensor

Unlike traditional fluorophores that typically experience fluorescence quenching upon aggregation, aggregation-induced emission (AIE)-active molecules exhibit a unique behavior of enhanced fluorescence emission when they aggregate, resulting from constrained intramolecular rotations upon interaction with biomolecules or during aggregate formation [38]. While this inherent characteristic makes AIE probes versatile "turn-on" sensing platforms for DNA and proteins, their interactions with other interfering biomolecules may compromise assay selectivity. The combination of AIE probes with graphene oxide (GO) has spurred the creation of innovative AIE-based biosensors known for their enhanced selectivity, and recent progress in integrating AIE probes, GO, and single-stranded DNA (ssDNA) has resulted in the development of rapid "turn-on" fluorescence methods for detecting DNA sequences with complementary counterparts [39]. These advancements highlight the potential of GO-assisted AIE biosensors as promising platforms for the accurate and reliable detection of biomolecules.

Our research team has created a GO-assisted AIE biosensor tailored for the detection of charged collagen peptides (Fig. 8.7A) [40]. Through the strategic conjugation of the AIE-active fluorophore TPE with peptide $G(PRGPOG)_5$, a novel triple-helical AIE-peptide probe, TPE-PRG, is devised to specifically target peptide EOG with the amino acid sequence $G(EOG)_{10}$, representing a prototypical example of charged amino acid-rich collagen fragments (Fig. 8.7A). The triple-helical peptide probe TPE-PRG displays pronounced fluorescence, attributed to the constrained intramolecular rotation of TPE within its trimeric conformation, while its interaction with GO leads to efficient fluorescence quenching. The introduction of the target collagen peptide EOG will prompt the release of the probe from the GO surface, resulting in the restoration of fluorescence. The resulting TPE-PRG/GO complex establishes an intricately designed 'turn-on' sensing platform with exceptional specificity for the target collagen peptide (Fig. 8.7A).

The peptide probe TPE-PRG displayed strong fluorescence at 472 nm, but upon the addition of GO, its fluorescence gradually diminished, reaching near-extinction at a GO concentration of 12 μg/mL, underscoring the exceptional quenching efficiency of GO and its potential for heightened sensitivity (Fig. 8.7B). Through subjecting the

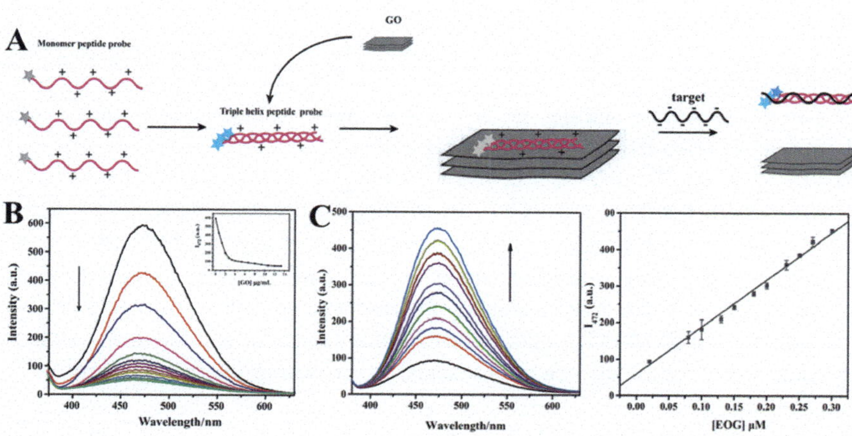

Fig. 8.7 **A** Schematic illustration of the concept of using a triple helical AIE-peptide probe and GO platform to detect target collagen sequences. **B** The fluorescence quenching of the triple helical probe peptide TPE-PRG by various concentrations of GO. **C** The linear fluorescence restoration of the TPE-PRG/GO platform by the target peptide EOG [40]. Reprinted from ref. [40], Copyright © 2020, with permission from Royal Society of Chemistry

TPE-PRG probe to hybridization with diverse concentrations of the target peptide EOG, a linear correlation emerged between the concentration of EOG and the fluorescence intensity of TPE-PRG (Fig. 8.7C). Remarkably, this TPE-PRG/GO sensing platform showcased an extensive linear range spanning from 20 to 300 nM, facilitating precise detection of peptide EOG at a low concentration of 17 nM (Fig. 8.7C). The TPE-PRG/GO biosensor provides a robust means of detecting charged collagen peptides, enhancing our capability to investigate collagen degradation dynamics and the pathology of collagen-related diseases.

8.5.3 WS_2 and MoS_2 Biosensing Platforms for Collagen Peptides

Transition metal dichalcogenides, such as WS_2 and MoS_2, are a class of recently uncovered two-dimensional layered nanomaterials akin to graphene, garnering significant interest for their distinct optical and catalytic attributes [41]. Unlike graphene, transition metal dichalcogenides nanosheets can be synthesized without oxidation treatment, offering controlled performance, and they can also incorporate thiolated compounds through simple self-assembly processes. The interaction between dye-labeled ssDNA and MoS_2 induces fluorescence quenching, while hybridization with its complementary target DNA sequence leads to fluorescence recovery, enabling precise assessment of the target DNA sequence [42]. Despite their potential, these MoS_2-based detection platforms have primarily utilized ssDNA as probe molecules, leaving the interaction between transition metal dichalcogenides nanosheets and other biomolecules such as amino acids, peptides, and proteins relatively unexplored.

Our research group presents a novel WS_2/MoS_2-based biosensing platform that incorporates an Arg-rich probe peptide FAM-PRG (with sequence FAM-G(PRGPOG)$_5$), where binding to WS_2/MoS_2 induces fluorescence quenching, and subsequent interaction with the target collagen peptide POG (with sequence G(POG)$_{10}$) leads to fluorescence recovery, enabling specific peptide POG detection (Fig. 8.8A) [43]. Peptide FAM-POG demonstrated significant fluorescence, partly quenched by WS_2, while the prominent fluorescence of peptide FAM-PRG experienced nearly complete quenching upon WS_2 interaction (Fig. 8.8B, C). MoS_2 nanosheets similarly displayed enhanced fluorescence quenching of peptide FAM-PRG compared to FAM-POG, underscoring the influence of peptide design, which incorporates five additional Arg amino acids, in promoting interactions with WS_2/MoS_2 nanosheets.

The potential of the WS_2/FAM-PRG platform as a quantitative assay for peptide POG was further explored, wherein the hybridization of probe peptide FAM-PRG with varying concentrations of the target collagen peptide POG revealed a linear correlation between POG concentration and platform fluorescence intensity (Fig. 8.8D, E). This sensing platform spans a broad linear range from 25 to 400 nM,

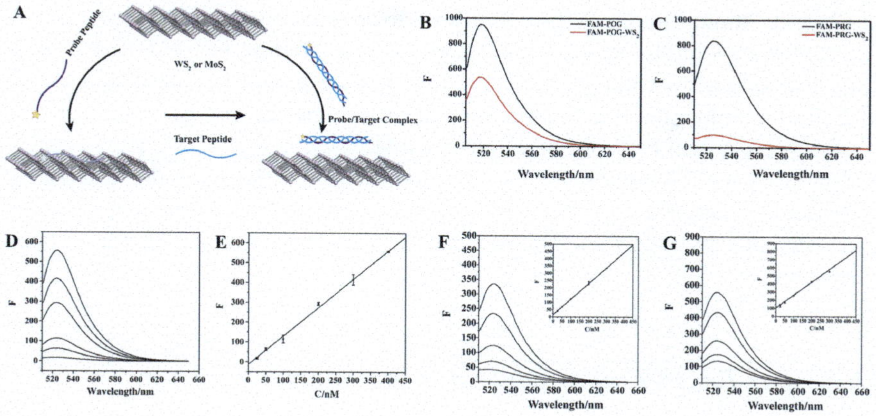

Fig. 8.8 **A** Schematic illustration of novel WS_2- and MoS_2- based biosensing platforms using peptide probe to specifically recognize the target. The fuorescence quenching of FAM-POG (**B**) and FAM-PRG (**C**) in the presence (red) and absence (black) of WS_2. (**D** and **E**) The proportional fluorescence restoration of the probe peptide FAM-PRG by the hybridization with target collagen peptide POG in PBS buffer. The linear fluorescence restoration of the probe peptide FAM-PRG by the hybridization with target collagen peptide POG in urine (**F**) and saliva (**G**) samples [43]. Reprinted from ref. [43], Copyright © 2017, with permission from Springer Nature.

allowing precise determination of the target collagen sequence POG even at a low concentration of 8 nM (Fig. 8.8D, E). The WS_2/FAM-PRG platform underwent additional evaluation with human urine and saliva samples, revealing a wide linear range of 25–300 nM for both sample types, and achieving detection limits of 41 and 21 nM, respectively (Fig. 8.8F, G). The WS_2-based platform provides a sensitive fluorescence assay with exceptional specificity for the target collagen peptide, facilitating accurate quantitative detection of collagen biomarkers in intricate biological fluids. The innovative establishment of WS_2- and MoS_2-based biosensors using non-ssDNA probes opens up exciting avenues for the creation of versatile biosensing platforms, poised to make a significant impact across a spectrum of biomedical applications.

8.6 Protocols for Fluorescence Assays in Collagen Peptide Detection

8.6.1 Fluorescence Self-quenching Assay Protocol

1. Peptide Preparation: Prepare a solution of 300 μM FAM-modified peptide probe (F-P) in 20 mM PBS buffer at pH 7.4. Similarly, prepare solutions of the target collagen peptides with varying concentrations (0, 15, 30, 60, 90, 120, 150, 300, 600, 900, 1200 μM) in 20 mM PBS buffer at pH 7.4.

2. Hybridization: Mix the F-P solution with the target collagen peptide solution in 20 mM PBS buffer at pH 7.4. Heat the peptide mixture to 90 °C for 20 min, followed by slow cooling at 4 °C for 24 h to allow complete hybridization.
3. Dilution: Dilute the F-P and target collagen peptide mixture by a factor of 300 to achieve a final F-P concentration of 1 μM.
4. Fluorescence Measurement: Record fluorescence spectra of the mixture at various molar ratios of F-P and the target collagen peptide. Measure the fluorescence intensity at 524 nm. Perform each measurement in triplicate for accurate results.

8.6.2 GO-Based Fluorescence Assay Protocol

1. Peptide Preparation: Prepare a solution of 300 μM FAM-modified peptide probe (F-P) in 20 mM PBS buffer at pH 7.4. Similarly, prepare solutions of the target peptide with varied concentrations (0, 15, 30, 60, 90, 120, 150, 300 μM) in 20 mM PBS buffer at pH 7.4.
2. Hybridization: Mix the F-P solution with the target collagen peptide solution in 20 mM PBS buffer at pH 7.4. Heat the peptide mixture to 90 °C for 20 min, followed by slow cooling at 4 °C for 24 h to allow complete hybridization.
3. Dilution: Dilute the F-P and target collagen peptide mixture by a factor of 300 to achieve a final F-P concentration of 1 μM.
4. GO Addition: Introduce a solution of GO with a final concentration of 3 μg/mL to the mixture for fluorescence measurements.
5. Fluorescence Measurement: Record fluorescence spectra of the mixture using appropriate excitation and emission wavelengths. Repeat all measurements three times to ensure accuracy and reliability.

References

1. Yan F, Fan K, Bai Z et al (2017) Fluorescein applications as fluorescent probes for the detection of analytes. TrAC, Trends Anal Chem 97:15–35
2. Han HH, Tian H, Zang Y et al (2021) Small-molecule fluorescence-based probes for interrogating major organ diseases. Chem Soc Rev 50(17):9391–9429
3. Pan S, Ding A, Li Y et al (2023) Small-molecule probes from bench to bedside: advancing molecular analysis of drug–target interactions toward precision medicine. Chem Soc Rev 52(16):5706–5743
4. Chinen AB, Guan CM, Ferrer JR et al (2015) Nanoparticle probes for the detection of cancer biomarkers, cells, and tissues by fluorescence. Chem Rev 115(19):10530–10574
5. Ueno T, Nagano T (2011) Fluorescent probes for sensing and imaging. Nat Methods 8:642–645
6. Liu X, Liu Y, Liu Q (2023) Fluorescent sensing platforms for detecting and imaging the biomarkers of Alzheimer's disease. Biosensors 13:515
7. Fang X, Zheng Y, Duan Y et al (2018) Recent advances in design of fluorescence-based assays for high-throughput screening. Anal Chem 91(1):482–504

8. Yao J, Yang M, Duan Y (2014) Chemistry, biology, and medicine of fluorescent nanomaterials and related systems: new insights into biosensing, bioimaging, genomics, diagnostics, and therapy. Chem Rev 114(12):6130–6178
9. Birmingham JD, Vilim V, Kraus VB (2006) Collagen biomarkers for arthritis applications. Biomarker Insights 1:117727190600100020
10. Liu T, Wang X, Karsdal MA et al (2012) Molecular serum markers of liver fibrosis. Biomarker insights 7: BMI. S10009
11. Willumsen N, Bager C, Karsdal MA (2019) Matrix metalloprotease generated fragments of type VI collagen have serum biomarker potential in cancer–a proof of concept study. Translational Oncol 12(5):693–698
12. López B, González A, Díez J (2010) Circulating biomarkers of collagen metabolism in cardiac diseases. Circulation 121(14):1645–1654
13. Dj L, He Y, Ss V et al (2011) A novel marker for assessment of liver matrix remodeling: an enzyme-linked immunosorbent assay (ELISA) detecting a MMP generated type I collagen neo-epitope (C1M). Biomarkers 16:616–628
14. Barascuk N, Veidal SS, Larsen L et al (2010) A novel assay for extracellular matrix remodeling associated with liver fibrosis: an enzyme-linked immunosorbent assay (ELISA) for a MMP-9 proteolytically revealed neo-epitope of type III collagen. Clin Biochem 43:899–904
15. Bay-Jensen A-C, Liu Q, Byrjalsen I et al (2011) Enzyme-linked immunosorbent assay (ELISAs) for metalloproteinase derived type II collagen neoepitope, CIIM—increased serum CIIM in subjects with severe radiographic osteoarthritis. Clin Biochem 44:423–429
16. Bager CL, Willumsen N, Christiansen C et al (2019) Bone and soft tissue turnover in relation to all-cause mortality in postmenopausal women. J Gerontol: Ser A 74(7):1098–1104
17. Srinivas GR, Barrach HJ, Chichester CO (1993) Quantitative immunoassays for type II collagen and its cyanogen bromide peptides. J Immunol Methods 159:53–62
18. Pernasetti F, Nickel J, Clark D et al (2006) Novel anti-denatured collagen humanized antibody D93 inhibits angiogenesis and tumor growth: an extracellular matrix-based therapeutic approach. Int J Oncol.
19. Mueller J, Gaertner FC, Blechert B et al (2009) Targeting of tumor blood vessels: a phage-displayed tumor-homing peptide specifically binds to matrix metalloproteinase-2-processed collagen IV and blocks angiogenesis in vivo. Mol Cancer Res 7:1078–1085
20. Demchenko AP, Demchenko AP (2015) Fluorescence detection techniques. In: Introduction to fluorescence sensing, pp 69–132
21. Su Q, Jiang C, Gou D et al (2021) Surface plasmon-assisted fluorescence enhancing and quenching: from theory to application. ACS Appl Bio Mater 4(6):4684–4705
22. Sapsford KE, Berti L, Medintz IL (2006) Materials for fluorescence resonance energy transfer analysis: beyond traditional donor–acceptor combinations. Angew Chem Int Ed 45(28):4562–4589
23. Stoddart LA, White CW, Nguyen K et al (2016) Fluorescence-and bioluminescence-based approaches to study GPCR ligand binding. Br J Pharmacol 173(20):3028–3037
24. Liao J, Madahar V, Dang R et al (2021) Quantitative FRET (qFRET) technology for the determination of protein–protein interaction affinity in solution. Molecules 26(21):6339
25. Shi L, Rosenzweig N, Rosenzweig Z (2007) Luminescent quantum dots fluorescence resonance energy transfer-based probes for enzymatic activity and enzyme inhibitors. Anal Chem 79(1):208–214
26. Jozawa H, Kabir MG, Zako T et al (2016) Amorphous protein aggregation monitored using fluorescence self-quenching. FEBS Lett 590(20):3501–3509
27. Quinn SD, Dalgarno PA, Cameron RT et al (2014) Real-time probing of β-amyloid self-assembly and inhibition using fluorescence self-quenching between neighbouring dyes. Mol BioSyst 10(1):34–44
28. Sun X, Fan J, Li X et al (2016) Colorimetric and fluorometric monitoring of the helix composition of collagen-like peptides at the nM level. Chem Commun 52:3107–3110
29. Nian L, Hu Y, Fu C et al (2018) Fluorescence self-quenching assay for the detection of target collagen sequences using a short probe peptide. Talanta 176:492–498

30. Sun X, Yao L, Fu C et al (2019) Detection of target collagen peptides with single amino acid mutation using two fluorescent peptide probes. J Mater Chem B 7:7676–7682
31. Zhu Y, Murali S, Cai W et al (2010) Graphene and graphene oxide: synthesis, properties, and applications. Adv Mater 22:3906–3924
32. Xiao X, Zhang Y, Zhou L et al (2022) Photoluminescence and fluorescence quenching of graphene oxide: a review. Nanomaterials 12(14):2444
33. Li D, Zhang W, Yu X et al (2016) When biomolecules meet graphene: from molecular level interactions to material design and applications. Nanoscale 8(47):19491–19509
34. Chen L, Song L, Zhang Y et al (2016) Nitrogen and sulfur codoped reduced graphene oxide as a general platform for rapid and sensitive fluorescent detection of biological species. ACS Appl Mater Interfaces 8:11255–11261
35. Sun X, Fan J, Ye W et al (2016) A highly specific graphene platform for sensing collagen triple helix. J Mater Chem B 4:1064–1069
36. Ge S, Zhang L, Zhang Y et al (2017) Nanomaterials-modified cellulose paper as a platform for biosensing applications. Nanoscale 9(13):4366–4382
37. Cai X, Yang Q, Ding J et al (2016) Paper-based FRET for the direct detection of collagen triple helix. J Mater Chem B 4:7009–7013
38. Hong Y, Lam JWY, Tang BZ (2011) Aggregation-induced emission. Chem Soc Rev 40(11):5361–5388
39. Tyagi A, Chu KL, Abidi IH et al (2017) Single-probe multistate detection of DNA via aggregation-induced emission on a graphene oxide platform. Acta Biomater 50:334–343
40. Sun X, Qiao Y, Li W et al (2020) A graphene oxide-aided triple helical aggregation-induced emission biosensor for highly specific detection of charged collagen peptides. J Mater Chem B 8:6027–6033
41. Chhowalla M, Shin HS, Eda G et al (2013) The chemistry of two-dimensional layered transition metal dichalcogenide nanosheets. Nat Chem 5(4):263–275
42. Zou R, Wang S, Chen C et al (2021) An enzyme-free DNA circuit-assisted MoS_2 nanosheet enhanced fluorescence assay for label-free DNA detection. Talanta 222:121505
43. Sun X, Fan J, Fu C et al (2017) WS_2 and MoS_2 biosensing platforms using peptides as probe biomolecules. Sci Rep 7:10290

Chapter 9
Collagen Mimetic Peptides for Targeting Pathological Collagen

Abstract The dysregulation of collagen remodeling, implicated in diseases like fibrosis, tumors, and arthritis, highlights the need for accurate detection of pathological collagen for diagnosis and prognosis. Innovative collagen mimetic peptides selectively bind to denatured collagen, promoting triple helix refolding by interacting with its unfolded chains, demonstrating high specificity for pathological collagen. This chapter explores the development of collagen mimetic pathological collagen targeting peptide probes (PCTPs) as crucial tools for targeted collagen recognition. It delves into various triple helical peptide probes featuring repetitive Pro-Hyp-Gly and Gly-Hyp-Hyp sequences, alongside variants incorporating aromatic residues Tyr and Phe at the X position and photo-triggered Gly. Additionally, it underscores the construction of single-stranded peptides using bulky dye and repulsion-destabilized methods, with a particular focus on an intrinsically non-trimerizing peptide probe designed for targeting pathological collagen. Furthermore, double-stranded peptides, including dimeric collagen mimic peptides and cyclic collagen mimetic peptides, are introduced. The chapter concludes with detailed protocols for targeting pathological collagen, encompassing detection in solutions, imaging in tissues, 3D cells and in vivo applications.

Keywords Pathological collagen · Triple helical peptide probes · Single-stranded · Double-stranded · Imaging

9.1 Introduction to Collagen Mimetic Peptides for Targeting Pathological Collagen

Collagen, the most abundant protein within the human body, plays a foundational role by acting as a molecular scaffold that provides structural integrity to diverse tissues and orchestrates a range of vital physiological processes, including tissue regeneration, cell proliferation, and wound healing [1–4]. Dysregulation of collagen remodeling has long been implicated as a central driver of critical diseases, such as fibrosis, tumors, and arthritis [5–7]. The abnormal accumulation of collagen is

© Springer Nature Singapore Pte Ltd. 2024
J. Xiao, *Collagen Mimetic Peptides and Their Biophysical Characterization*, Springer Series in Biomaterials Science and Engineering 20,
https://doi.org/10.1007/978-981-10-1941-8_9

a hallmark of fibrosis-related conditions, leading to the disruption of normal tissue architecture and impairing organ function [8–11]. Moreover, the anomalous degradation of diverse collagen types has been notably linked to tumor proliferation, invasion, and metastasis. Additionally, the excessive degradation of Type II collagen, a principal component of articular cartilage, is recognized as a pivotal factor in the onset of rheumatoid arthritis and osteoarthritis [12–14]. Hence, the specific and sensitive detection of pathological collagen assumes a pivotal role in disease diagnosis, prognosis, and the formulation of therapeutic strategies for these collagen-related disorders [15].

Over the past decades, considerable endeavors have been focused on engineering antibodies and peptide probes to achieve precise collagen binding [16–18]. ELISA-based screening revealed the monoclonal antibodies E1E5 and E4A11, demonstrating distinct recognition of type II collagen [19–21]. Additionally, exploration of collagen-binding proteins like von Willebrand factor (vWF) has unveiled peptides such as WREPSFCALS, TKKTLRT, and LRELHLNNN, all targeting type I collagen [22–24]. Phage display libraries have led to the discovery of collagen-specific peptides, such as KLWVLPK for type IV collagen in tumors and WYRGRL for type II collagen in articular cartilage [25, 26]. However, these assays confront substantial obstacles owing to collagen's intricate structure, leading to restricted specificity, compromised binding affinity, and notably, an inability to distinguish denatured collagen from its intact counterpart.

The concept of denatured collagen-specific recognition has ignited increased attention, propelled by the emergence of a novel class of collagen mimetic peptides featuring repetitive (Gly-Pro-Hyp)$_n$ sequences. Diverging from previously documented peptide probes, these innovative collagen mimetic peptides exhibit conformation-dependent binding to denatured collagen, engaging with unfolded chains of pathological collagen to undergo triple helix refolding (Fig. 9.1) [27]. This unique mechanism imparts high specificity for pathological collagen, excluding interaction with native collagen that lacks the unfolded region required for binding these mimetic peptides. Emerging as pivotal tools for targeted collagen recognition, the novel pathological collagen targeting peptide probes (PCTPs) have found widespread application in precise in vitro and in vivo detection of pathological collagen across diverse conditions, including tumors, arthritis, and fibrosis. Significant strides have been achieved in the development of a range of collagen mimetic peptides, encompassing triple helical PCTPs, single-stranded PCTPs, and double-stranded PCTPs, with the aim of enhancing binding affinity and overall assay effectiveness.

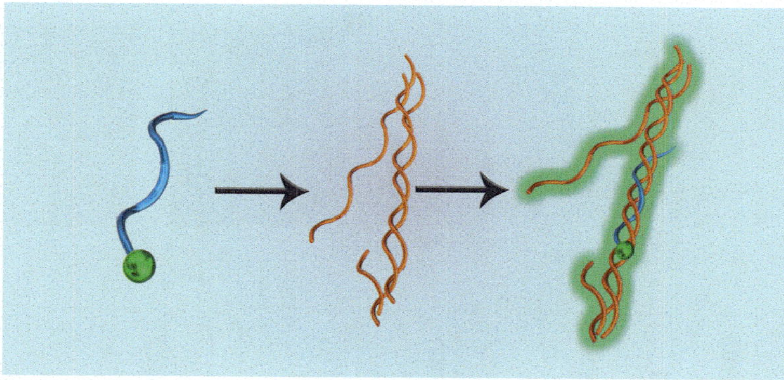

Fig. 9.1 Schematic diagram of PCTP binding to pathological collagen

9.2 Triple Helical Peptides for Targeting Pathological Collagen

9.2.1 (Pro-Hyp-Gly)$_n$ Sequence for Identifying Pathological Collagen

Yu and colleagues discovered that the (Pro-Hyp-Gly)$_n$ sequence could selectively target pathological collagen but not intact collagen [28]. Two carboxyfluorescein (FAM) labeled peptides (Pro-Hyp-Gly)$_6$ and (Pro-Hyp-Gly)$_{10}$ were synthesized, and they exhibited negligible binding affinity towards both intact and pathological collagen when in the triple helical conformation. Upon preheating above their respective melting temperatures (T_m), the two peptides unfolded from their triple helical structures into single-stranded states, demonstrating a strong affinity towards pathological collagen (Fig. 9.2A). The postulation was that the (Pro-Hyp-Gly)$_n$ sequence could potentially bind to pathological collagen by hybridizing with the unfolded chains of collagen molecules.

A new modification technique was explored by conjugating the (Pro-Hyp-Gly)$_7$ sequence with poly(ethylene glycol) (PEG) to control the adhesiveness of collagen to cells (Fig. 9.2B). When the collagen film was treated with the peptide-conjugated polymer, the resulting film was nearly devoid of fibroblasts, whereas the untreated film exhibited a homogeneous distribution of fibroblasts. These findings suggested that the modified collagen films had reduced cell adhesion, indicating that the peptide-conjugated polymer successfully altered the adhesive properties of the collagen. By exploiting the (Pro-Hyp-Gly)$_n$ sequence's binding capability to pathological collagen, this modification technique has the capacity to regulate cell organization in collagen matrices and open up novel possibilities for tissue engineering applications.

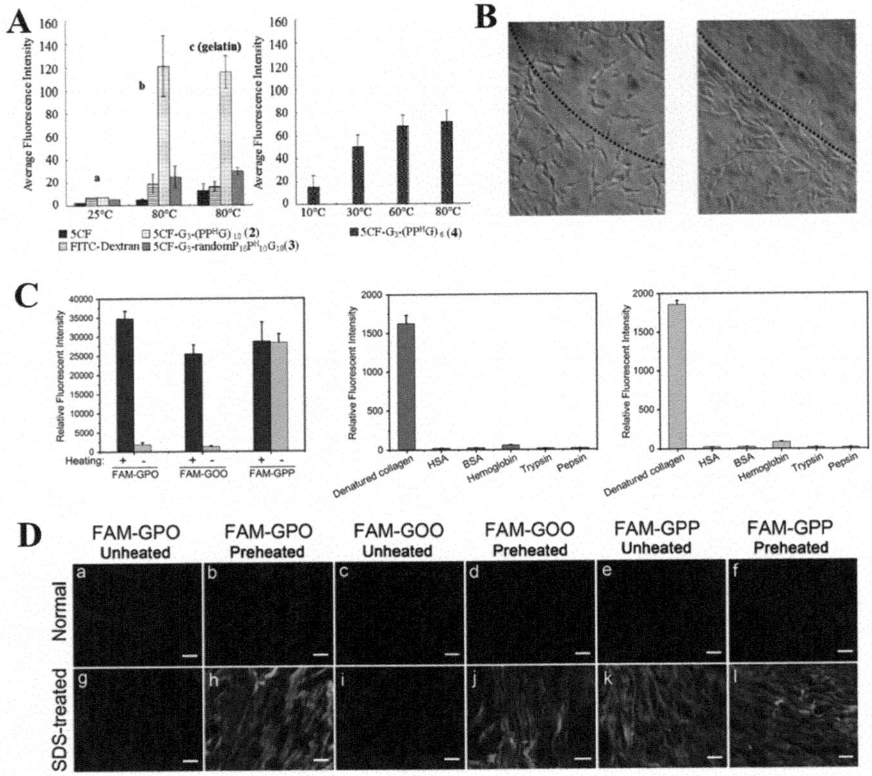

Fig. 9.2 **A** Fluorescence intensities of collagen films (type I, bovine) treated with 5CF-labeled CMPs and other control samples. **B** Optical micrographs of human fibroblasts cultured on collagen films that were pretreated with mPEG2000 or mPEG2000-$(POG)_7$ [28] (Reprinted from ref. [28], Copyright © 2005, with permission from American Chemical Society). **C** Binding affinity of peptide probes FAM-GPO, FAM-GOO and FAM-GPP toward gelatin films in a 96-well plate (left). Peptide solutions were preheated at 80 °C before their addition into the wells in the "Heating +" group, while they were stored at 4 °C and directly used in the "Heating −" group. Binding affinity of preheated FAM-GOO (middle) and unheated FAM-GPP (right) with denatured type I collagen and other proteins (HSA, BSA, hemoglobin, trypsin, and pepsin). **D** Fluorescence micrographs of normal and SDS-treated rat tendon tissues stained with unheated and preheated FAM-GPO, FAM-GOO, and FAM-GPP [27] (Reprinted from ref. [27], Copyright © 2020, with permission from Royal Society of Chemistry)

9.2.2 Peptide Probes with the Repetitive Gly-Hyp-Hyp and Gly-Pro-Pro Sequences

Our research team has developed two innovative collagen-targeting peptide probes: FAM-GOO (with the sequence FAM-$(Gly\text{-}Hyp\text{-}Hyp)_8$-Gly) and FAM-GPP (with the sequence FAM-$(Gly\text{-}Pro\text{-}Pro)_8$-Gly) [27]. FAM-GOO could form a stable triple helical structure with a T_m of 49 °C, whereas FAM-GPP remained in a single-stranded conformation even at a low temperature of 4 °C. Similar to FAM-GPO, FAM-GOO necessitated preheating for binding with gelatin, while both unheated and preheated FAM-GPP exhibited the capability to bind with gelatin (Fig. 9.2C). Preheated FAM-GOO and unheated FAM-GPP demonstrated exceptional specificity towards denatured collagen, while displaying no detectable cross-reactivity with other proteins, including human serum albumin (HSA), bovine serum albumin (BSA), hemoglobin, trypsin, and pepsin (Fig. 9.2C).

The tissue staining efficacy of the three peptide probes FAM-GPO, FAM-GOO and FAM-GPP was evaluated (Fig. 9.2D). In normal tissues, regardless of whether they were preheated or not, fluorescence micrographs consistently showed no staining by any of the three peptide probes, indicating their incapacity to bind to intact collagen. In SDS-treated tissues, the fluorescence micrographs revealed a distinct pattern: both FAM-GPO and preheated FAM-GOO effectively identified denatured collagen when preheated but not when unheated, whereas both preheated and unheated FAM-GPP consistently displayed strong staining of denatured collagen. FAM-GOO and FAM-GPP exhibited comparable effectiveness in selectively staining denatured collagen in tissues, with the noteworthy advantage that FAM-GPP's unique single-stranded conformation eliminated the requirement for thermal pretreatment (Fig. 9.2D). These peptide probes, derived from natural amino acids, presented a convenient, biocompatible, and highly specific biosensing approach for pathological collagen, showcasing significant promise for the diagnosis and treatment of collagen-related disorders.

9.2.3 *Peptide Probes with Aromatic Residues Tyr and Phe at the X Position*

Our research group has created a novel series of PCTP probes, designated as FAM-GYO, FAM-GFO, FAM-GDO, and FAM-GAO, by introducing non-imino acids at the X position of the FAM-$(GPO)_3GXO(GPO)_4$ sequence, enabling the selective detection of pathological collagen [29]. The inclusion of aromatic residues Tyr and Phe at the X position did not significantly alter the binding affinity of the peptide probes compared to Pro, whereas the binding affinity notably diminished with the incorporation of charged residue Asp and hydrophobic Ala (Fig. 9.3A). Moreover, FAM-2GYO and FAM-2GAO were engineered to feature Tyr and Ala at two X positions within the FAM-$(GPO)_2GXOGPOGXO(GPO)_3$ sequence, demonstrating a significant impact on the binding affinity of the peptide probes upon substituting Pro with Tyr or Ala at the two X positions (Fig. 9.3A).

Similar to FAM-GPO, both FAM-GYO and FAM-GFO showed minimal binding affinity with all other proteins, including pepsin, trypsin, and hemoglobin, except for denatured type I collagen, underscoring the high specificity of these peptide probes in detecting denatured collagen (Fig. 9.3B). Fluorescence micrographs of FFPE tissues from liver fibrosis, liver cancer, rectal cancer, and esophageal cancer stained with FAM-GYO and FAM-GFO displayed strong green fluorescence, indicating their

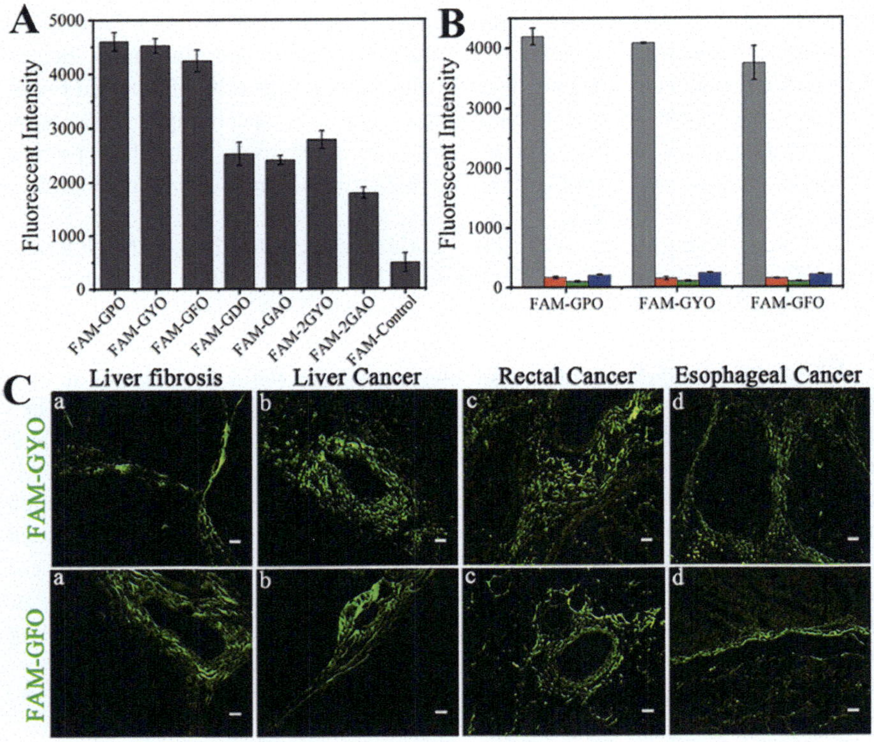

Fig. 9.3 **A** Binding affinity of peptide probes FAM-GPO, FAM-GYO, FAM-GFO, FAM-GDO, FAMGAO, FAM-2GYO, FAM-2GAO, and FAM-control toward gelatin films in a 96-well plate. **B** Binding affinity of FAM-GPO, FAM-GYO, and FAM-GFO with denatured type I collagen (gray), pepsin (red), trypsin (green), and hemoglobin (blue), respectively. **C** Fluorescence micrographs of liver fibrosis, liver cancer, rectal cancer, and esophageal cancer; tissue sections stained with FAM-GYO and FAM-GFO [29]. Reprinted from ref. [29], Copyright © 2020, with permission from American Chemical Society

broad-spectrum biosensing capabilities similar to FAM-GPO for detecting denatured collagen (Fig. 9.3C). These findings suggest that aromatic interactions could enhance PCTP binding to pathological collagen, and the inclusion of nonimino acids, especially aromatic residues at the X and Y positions, may represent a promising strategy for designing potent PCTP probes.

9.2.4 Photo-Triggered Peptide Probe for Discerning Pathological Collagen

The introduction of the photo-cleavable cage group nitrobenzyl (NB) into the (Gly-Pro-Hyp)$_n$ sequences led to the creation of photo-triggered peptide probes designed to target pathological collagen [30]. Conjugating NB to the central Gly of (Gly-Pro-Hyp)$_9$ resulted in the production of the peptide probe CF^{NB}(GPO)$_9$, characterized by the sequence CF-Gly$_3$-(GlyProHyp)$_4$-^{NB}GlyProHyp-(GlyProHyp)$_4$. Thermal unfolding curves indicated that CF^{NB}(GPO)$_9$ maintained a monomer conformation due to the presence of the bulky NB group, but upon short UV exposure, the NB group was removed, leading to the immediate formation of a stable triple helix. Microplate assays demonstrated that CF^{NB}(GPO)$_9$ (UV−) exhibited minimal binding affinity to both intact and pathological collagen, whereas photo-activated CF^{NB}(GPO)$_9$ (UV+) displayed strong affinity specifically towards pathological collagen. This highlighted UV as a convenient modulator for activating CF^{NB}(GPO)$_9$ for the selective recognition of pathological collagen.

The near-infrared fluorophore (NIRF) labelled ^{NB}(GPO)$_9$ peptide was explored to image pathological tissues with elevated MMP activity. After rapid UV-induced decaging, the NIRF-labeled peptide probe was promptly administered via the tail vein to mice with subcutaneous PC-3 prostate tumor xenografts. Serial in vivo fluorescence imaging sessions revealed the efficient penetration of the peptide probe through tumor vasculature and its accumulation at tumor sites. In contrast, the control peptide sequence devoid of triple-helix folding capacity exhibited minimal accumulation at the tumor site. Moreover, co-injection of a fluorescent beacon for MMP activity demonstrated the colocalization of MMP activity and peptide probe binding within tumors. The photo-triggered peptide probes present a robust approach for targeting a wide range of pathological conditions associated with collagen remodeling and heightened MMP activity.

9.3 Single Stranded Peptides for Targeting Pathological Collagen

9.3.1 A Single Stranded Fluorescence Peptide Probe with Bulky Dye for Detecting Collagen

Our research group has discovered a single stranded fluorescent peptide probe, ssFCMP, featuring the sequence of Ac-(Gly-Pro-Hyp)$_3$-Gly-Pro-Amp(FAM)-(Gly-Pro-Hyp)$_3$-Gly-NH$_2$ [31]. Due to the structural similarity between Amp ((2S, 4S)-aminoproline) and Hyp, the peptide shares an identical main chain with the previous peptide probe (Gly-Pro-Hyp)$_7$-Gly, thereby endowing it with a collagen-targeting capability similar to that of (Gly-Pro-Hyp)$_7$-Gly. Meanwhile, the inclusion of FAM (5(6)-carboxyfluorescein) at the side-chain of an unnatural amino acid Amp in the middle region, as opposed to the N-terminus, is intended to impede the formation of the homotrimeric peptide. The linear decrease in CD intensity during melting for ssFCMP confirmed the maintenance of its single-stranded state (Fig. 9.4A).

Fluorescence micrographs of tendon sections stained with unheated ssFCMP exhibited strong green fluorescence, demonstrating ssFCMP's remarkable ability to

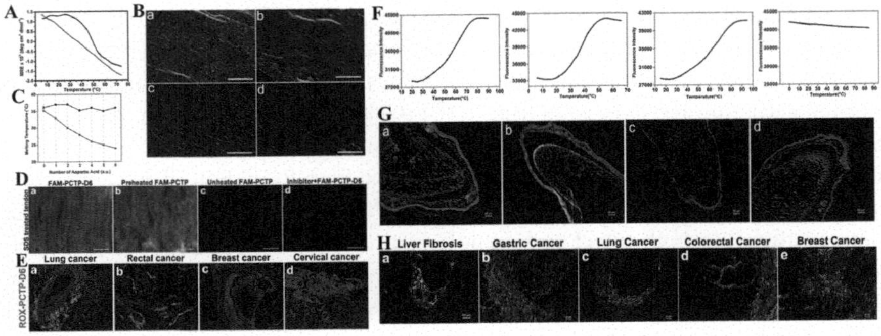

Fig. 9.4 **A** CD thermal unfolding of ssFCMP (red) and control (black) [31]. **B** Fluorescence micrographs of rat tendon tissue sections [31] (Reprinted from ref [31], Copyright © 2017, with permission from Royal Society of Chemistry). **C** The relationship between the melting temperature (T_m) of the PCTP series peptides with the number of Asp at pH 7.4 (red) and pH 2.7 (black) [32]. **D** Fluorescence micrographs of SDS-treated rat tendon and tissue sections [32]. **E** Cancer tissue staining using peptide probe ROX-PCTP-D6 [32] (Reprinted from ref. [32], Copyright © 2020, with permission from American Chemical Society). **F** Conformation of the peptide probes F-GPO-10, F-GPP-10, F-GOO-10, and F-GOP-10 [33]. **G** Specific recognition of pathological collagen by peptide probe F-GOP-10. **H** Histopathology staining using peptide probe F-GOP-10 [33] (Reprinted from ref. [33], Copyright © 2020, with permission from John Wiley & Sons, Inc.)

target collagen in tendon tissues (Fig. 9.4B). Conversely, fluorescence micrographs of tendon sections stained with heated FCMP, but not unheated FCMP, displayed strong green fluorescence, indicating the necessity of heating FCMP to attain the single-stranded state for efficient collagen imaging. FAM alone, used as a control, failed to stain collagen in tendon tissues, resulting in weak green fluorescence (Fig. 9.4B). Both unheated ssFCMP and preheated FCMP effectively stained collagen in tissues, with the distinct advantage that ssFCMP did not require heating as it consistently maintained the unfolded, single-stranded state. The ssFCMP offers a novel single-stranded fluorescent peptide probe for targeting collagen strands under physiological conditions, eliminating the need for thermal or ultraviolet treatment and reducing the risk of tissue damage.

9.3.2 In Situ Imaging of Pathological Collagen by Repulsion-Destabilized Peptide Probes

Our research group has established an electrostatic repulsion strategy to significantly reduce the triple helical stability of collagen-targeting peptides, resulting in the design of a series of PCTP probes (PCTP, PCTP-D1, PCTP-D2, PCTP-D3, PCTP-D4, PCTP-D5, and PCTP-D6) by introducing zero to six Aspartic acid (Asp) residues to the N-terminus of the $(Gly\text{-}Pro\text{-}Hyp)_7$ sequence [32]. CD spectroscopy analysis unveiled a gradual reduction in the stability of the triple helix structure with an escalating number of Asp residues, attributed to heightened electrostatic repulsion at pH7.4 (Fig. 9.4C). The least stable peptide, PCTP-D6, was selected as the collagen-targeting sequence for the development of two novel peptide probes, each featuring a distinct fluorescent dye at the N-terminus: FAM-PCTP-D6 and ROX-PCTP-D6. FAM-PCTP-D6 demonstrated the formation of substantial monomer species under physiological conditions (pH7.4) at room temperature, as confirmed through colorimetric and fluorescence assays.

Fluorescence micrographs of SDS-treated tendon tissue sections stained with FAM-PCTP-D6 showed strong green fluorescence, confirming its ability to detect denatured collagen, while control peptide probe FAM-PCTP exhibited fluorescence only when preheated (Fig. 9.4D). The inhibitory effect of $G(POG)_{10}$ on denatured collagen staining by FAM-PCTP-D6 suggests that this novel peptide probe operates similarly to previously reported probes, specifically recognizing denatured collagen by interacting with the unfolded chains (Fig. 9.4D). Notably, FAM-PCTP-D6 could be used directly for tissue staining without any pretreatment, and the additional aspartic acid (Asp) residues did not compromise its specificity for targeting denatured collagen. Furthermore, ROX-PCTP-D6 has been employed to specifically target pathological collagen in various cancer tissues, encompassing rectal cancer, lung cancer, cervical cancer, and breast cancer (Fig. 9.4E). This probe has proven its efficacy in targeting pathological collagen within the connective tissues of both animal disease models and human patients, serving as a powerful tool for histopathology staining.

9.3.3 An Intrinsically Non-trimerizing Peptide Probe for Recognizing Pathological Collagen

Our research group for the first time constructed an intrinsically nontrimerizing peptide probe, F-GOP-10 (sequence FAM-Ahx-$(Gly\text{-}Hyp\text{-}Pro)_{10}$-$NH_2$), for targeting pathological collagen, which involves a positional exchange of Pro and Hyp compared to peptide F-GPO-10 (sequence FAM-Ahx-$(Gly\text{-}Pro\text{-}Hyp)_{10}$-$NH_2$) [33]. Real-time monitoring of fluorescence intensity at 525 nm revealed that peptide probes F-GPO-10, F-GPP-10, and F-GOO-10 all exhibited a gradual increase in fluorescence with rising temperature until a plateau was reached, signifying the transition of the triple helix into the monomer; however, no melting transition was observed for peptide F-GOP-10, confirming its consistent monomer conformation (Fig. 9.4F). The rearrangement of Pro and Hyp positions within peptide probe F-GOP-10 likely eradicated the favorable ring conformations at the X and Y positions, resulting in the complete destabilization of the triple helical structure.

Fluorescence micrographs of normal ear sections stained with unheated F-GOP-10 showed no green fluorescence, indicating its inability to target intact collagen, while those of impaired ear sections stained with unheated F-GOP-10 displayed strong green fluorescence, demonstrating its high specificity for pathological collagen

(Fig. 9.4G). Fluorescence micrographs of impaired ear tissue sections stained with F-GOP-10 in the presence of inhibitor $G(POG)_{10}$ displayed very weak green fluorescence, suggesting that F-GOP-10 targets pathological collagen through hybridization with unfolded chains. The control peptide probe F-GOP-10-random, unable to stain pathological collagen in impaired ear tissues as indicated by weak green fluorescence, suggested that the specificity of F-GOP-10 toward denatured collagen was attributed to the $(GOP)_{10}$ sequence (Fig. 9.4G). F-GOP-10 has demonstrated its effectiveness in detecting pathological collagen across a range of tissues, encompassing conditions such as liver fibrosis, gastric cancer, lung cancer, colorectal cancer, and breast cancer (Fig. 9.4H). In contrast to the conventional Masson's trichrome staining method, F-GOP-10 exhibits remarkable selectivity, specifically binding to pathological collagen while avoiding interference with intact collagen. This monomeric peptide probe offers precise concentration determination and eliminates the need for potentially harmful pretreatment, making it highly promising for histopathology staining and various clinical applications.

9.4 Double Stranded Peptides for Targeting Pathological Collagen

9.4.1 Dimeric Collagen Mimetic Peptide for Enriching Collagen Fragments

While the single-stranded PCTP has demonstrated efficacy in detecting significant-length pathological collagen, it exhibits limited ability to capture and identify short collagen fragments. To address this limitation, Yu group has developed an end-tethered dimeric collagen mimetic peptide D-CHP (sequence $[Ac\text{-}(GPO)_6\text{-}G_3]_2$-K-GK), to enhance hybridization with sparse collagen fragments through a 1:1 stoichiometry [34]. The peptide was synthesized via Fmoc-mediated solid-phase peptide synthesis (SPPS), involving the incorporation of a parallel-protected Fmoc-Lys(Fmoc)-OH residue during the process, followed by simultaneous extension of the two GPO chains after the branch point. D-CHP displayed a distinctive CD profile associated with a triple helical structure, characterized by a melting transition occurring at 44 °C, notably 7 °C higher than that observed for the monomeric M-CHP (Fig. 9.5A). Additionally, D-CHP exhibited increased affinity for crosslinked gelatin compared to M-CHP, with a more pronounced difference observed at 25 °C (Fig. 9.5B). D-CHP likely forms a more stable complex with denatured collagen, potentially by adopting a hetero-triple helix structure composed of two tethered CHP strands.

The binding affinity of D-CHP and M-CHP to collagen fragments was assessed using an ELISA-like monolayer capture surface created by covalently immobilizing CHPs onto the surface of a 96-well assay plate (Fig. 9.5C). Four peptides were synthesized to mimic collagen fragments, selected from the triple helical domain of

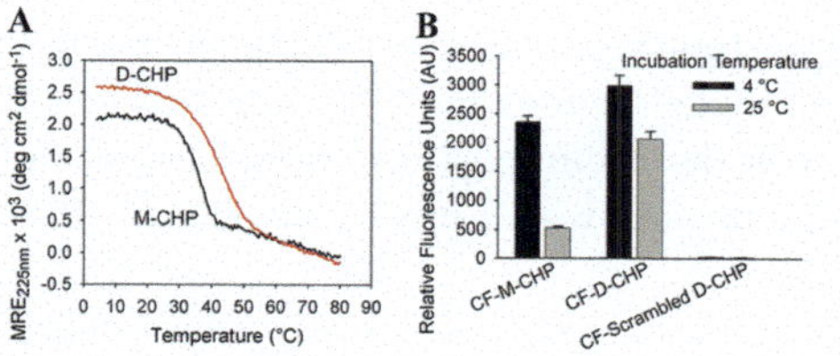

C

Synthetic Collagen Fragment	Sequence Location in type I collagen	K_D against surface immobilized D-CHP (nM)
GEO GPA GVQ GPO GPA GEE GK	438-457	139.2
GSO GPA GPK GSO GEA GRO GEA GLO GAK	501-527	10.8
GPD GKT GPO GPA GQD GRP GPA GPO GAR	537-563	61.0
GLT GPI GPO GPA GAO GDK	753-770	269.1

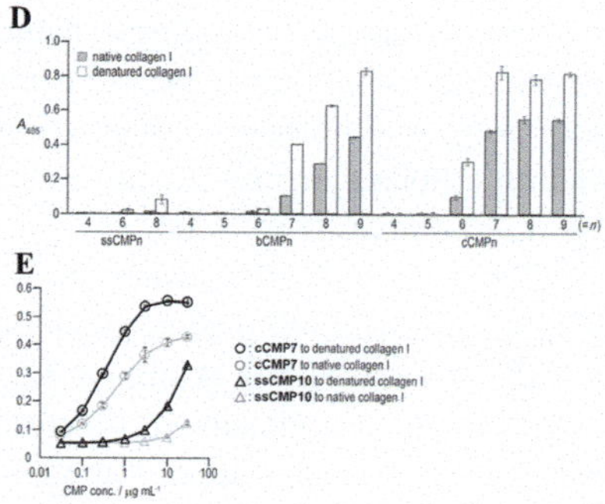

Fig. 9.5 **A** Triple helical folding and **B** gelatin binding of M- and D CHPs [24]. **C** Amino acid sequences of synthetic collagen fragments, their locations in Rat Col1a1, and K_D against surface-immobilized D-CHP [34] (Reprinted from ref. [34], Copyright © 2020, with permission from American Chemical Society). **D** Binding of 2 μg/mL CMPs with heating to native and heat-denatured collagen at 4 °C [35]. **E** Dose-dependent binding of cCMP7 and ssCMP10 with heating to native and heat-denatured collagen at 4 °C [35] (Reprinted from ref. [35], Copyright © 2018, with permission from John Wiley & Sons, Inc.)

the α_1 chain of rat type I collagen based on the criteria of lacking consecutive GPO repeats and covering various lengths and amino acid compositions. All synthetic collagen peptides demonstrated binding to D-CHP with K_D values spanning from 10 to 270 nM, while their interaction with M-CHP was minimal, emphasizing the pronounced advantage of D-CHP for capturing collagen fragments (Fig. 9.5C). Due to the synthetic collagen peptides' inherently low triple helical propensity, the presence of two tethered GPO strands in D-CHP enhances their capture by augmenting both the folding rate and the stability of the resulting triple helix. LC–MS/MS analysis of enriched mouse urine revealed an average of 383 collagenous peptide fragments per sample, a substantial increase compared to the 34 fragments detected in unenriched samples. D-CHP has proven highly effective in enriching collagen fragments from biological fluids, promising substantial contributions to peptidomics-based disease detection and biomarker discovery.

9.4.2 Cyclic Collagen Mimetic Peptide

Koide group has introduced a novel category of cyclic collagen mimetic peptides (cCMPs) comprising two parallel sequences of $(\text{Pro-Hyp-Gly})_n$ ($n = 4, 5, 6, 7, 8$, or 9) connected by a disulfide bond at both termini [35]. Their binding affinity to type I collagen was assessed through an enzyme-linked immunosorbent assay (ELISA) employing biotin-labeled cCMPs (Fig. 9.5D). The cCMP in the trimeric conformation exhibited minimal binding activity to native as well as pathological collagen, while the monomeric cCMP, following preheating treatment, demonstrated significant binding activity to both types of collagen. Remarkably, cCMPs displayed a strong preference for binding to pathological collagen over native collagen, possibly attributable to the higher prevalence of unfolded regions in pathological collagen, facilitating more effective interactions with the cyclic peptides. Furthermore, the binding efficiency of CMPs increased significantly with a greater number of (Pro-Hyp-Gly) repeats (Fig. 9.5D).

cCMP7 and ssCMP10 were chosen as representative designs, and their binding affinities toward both pathological and native collagen were compared (Fig. 9.5E). cCMP7 and ssCMP10 exhibited K_D values of 6.6×10^{-8} M and 9.5×10^{-6} M for pathological collagen, respectively, indicating that cCMP7 had an affinity for pathological collagen approximately two orders of magnitude higher than that of ssCMP10. Additionally, cCMP7 exhibited a K_D value of 1.1×10^{-7} M for native collagen, while the K_D value of ssCMP10 was below the detectable range (Fig. 9.5E). Molecular dynamics simulations indicated that cCMP forms a more stable hybridized product than its single-chain counterpart, potentially explaining its higher affinity for denatured collagen. The cyclic peptide cCMP7 offers a more robust probe for the specific recognition of collagen, showcasing significant potential in applications related to collagen-related diseases.

9.4.3 Protocols for Targeting Pathological Collagen

9.4.4 Protocol for Detecting Pathological Collagen in Solutions

1. Gelatin Preparation: Dissolve gelatin powder in 1 × PBS to create a 500 μg/mL gelatin solution. Maintain the gelatin solution at 70 °C using a water bath.
2. Coating the Microplate: Add approximately 50 μL of the prepared gelatin solution into each well of a 96-well microplate. Incubate the microplate at 4 °C for 12 h to allow the gelatin to coat the well surfaces.
3. Pre-targeting: Prepare a PBS solution containing 20 μM of carboxyfluorescein-labeled PCTP. The diluted sample solution was prepared at concentrations of 0.1-fold, 0.01-fold and 0.001-fold. Mix the PCTP solution and each sample solution in equal volumes and incubate the mixture at 4 °C for 4 h.
4. Peptide Binding: Add 50 μL of the mixture solution from the step 3 to each well coated with gelatin. Allow the binding between PCTP and gelatin to occur for 5 h at room temperature.
5. Washing: Wash the microplate three times with 1 × PBS at room temperature to remove unbound peptide.
6. Fluorescence Measurement: Measure the fluorescence signals using a microplate reader (e.g., SpectraMax M-2) with excitation at 485 nm and emission at 533 nm.
7. Data Analysis: Analyze the collected fluorescence data to quantify the binding of carboxyfluorescein-labeled PCTP to the gelatin-coated wells (Fig. 9.6).

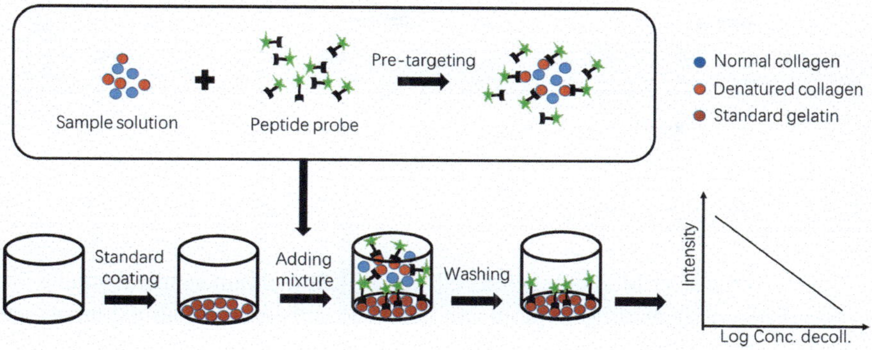

Fig. 9.6 Schematic illustration for the Detection of pathological collagen in solutions

9.4.5 Protocol for Imaging Pathological Collagen in Tissues

1. Blocking Buffer Preparation: Prepare 10 mL of blocking buffer by mixing 500 μL of goat serum, 1 mL of 10 × PBS, and 8.5 mL of deionized water.
2. Dilution Buffer Preparation: Prepare 1 mL of 10% (w/v) BSA solution in deionized water. Prepare 5 mL of PCTP dilution buffer by mixing 50 μL of 10% (w/v) BSA, 4.45 mL of deionized water and 500 μL of 10 × PBS.
3. PCTP solution: Dilute stock solutions of carboxyfluorescein-labeled PCTP to 10 μM using the dilution buffer. Use 100 μL of PCTP solution for staining each tissue section slide.
4. Tissue Preparation: Incubate the tissue slides in 1 × PBS buffer for 5 min at room temperature.
5. Blocking: Place tissue slides in a humidified chamber. Apply 0.5 mL of blocking solution to each slide. Incubate slides for 30 min at room temperature. Remove blocking solution by blotting slides on a paper towel.
6. PCTP Staining: Apply 100 μL of PCTP solution to cover the tissue sections on each slide. Incubate slides for 2 min to allow PCTP permeation. Cover tissue sections with Parafilm to prevent drying. Place slides in a humidified chamber. Incubate at 4 °C for 2 h.
7. DAPI Staining: Blot slides on a paper towel to remove excess PCTP. Prepare a 1:3000 dilution of DAPI solution in 1 × PBS. Apply approximately 100 μL of diluted DAPI solution to each tissue slide. Incubate for 1 min at room temperature.
8. Washing: Immerse slides in 1 × PBS buffer in a staining jar for 5 min. Repeat this washing step three times using fresh 1 × PBS to remove unbound staining agents.
9. Mounting: Add a drop of mounting medium to each tissue section. Cover with a glass cover slip, avoiding air bubbles. Place slides in a cardboard slide tray to protect from light.
10. Imaging: Image the tissue slides using a fluorescence microscope (Fig. 9.7).

Fig. 9.7 Schematic illustration for imaging pathological collagen in tissues

9.4.6 Protocol for Imaging Pathological Collagen in 3D Cells

1. Cell Maintenance: Maintain HeLa cells in DMEM medium supplemented with 10% FBS and penicillin/streptomycin. Incubate cells in a humidified incubator at 37 °C with 5% CO_2.
2. 3D Collagen Matrix Preparation: Dissolve type I collagen (4 mg/mL) in an acetic acid solution. Mix collagen solution with 0.34 N NaOH and 10 × PBS in an 8:1:1 volume ratio. Add 1×10^5 HeLa cells to the neutralized collagen solution. Thoroughly mix the cell–matrix mixture (250 μL/well). Pour the mixture onto a 24-well chambered cover glass. Incubate at 37 °C for 1 h to ensure gelation. Maintain 3D cells in DMEM medium supplemented with 10% FBS and penicillin/streptomycin for 24 h. Incubate cells in a humidified incubator at 37 °C with 5% CO_2. Fix the cell-embedded 3D collagen matrix with 4% paraformaldehyde in PBS for 0.5 h. Wash the constructs with PBS for 0.5 h.
3. PCTP Staining: Add 5 μM fluorescently labeled PCTP to the fixed cell–matrix constructs (500 μL/well). Stain the constructs without pre-blocking for 1 h at 4 °C. Perform four rounds of 30-min washes in PBS.
4. Nuclei Staining: Stain cell nuclei with Hoechst 33,258 for 30 min at room temperature.

9.4.7 Protocol for Imaging Pathological Collagen in Vivo

1. Peptide Solution Preparation: Prepare 100 μL of 20 μM PCTP solution for each mouse (20–30 g mouse weight).
2. Mouse Preparation: Position the mouse in a restrainer under a heated lamp. Disinfect the injection site (preferably the posterior part of the tail) with 75% alcohol.
3. Peptide Injection: Slowly withdraw the peptide solution into a 0.5 mL insulin syringe with a transparent barrel and a small gauge needle. Inject 100 μL of the PCTP solution into the tail vein.
4. Mouse Anesthetization: Put the mouse in an anesthesia induction chamber containing 2% isoflurane in oxygen. Verify the depth of anesthesia by monitoring mouse movements and respiration rate (~1/s).
5. Mouse Placement: Place the mouse on the imaging bed. Secure the mouse's muzzle inside the nose cone to maintain anesthesia (2% isoflurane in oxygen at 2 L/min) during imaging.

6. Imaging and Analysis: Capture NIR (excitation 685 nm, emission 720 nm) and white light photographs. Analyze the acquired NIR fluorescence images using software tools.

References

1. Xu S, Xu H, Wang W et al (2019) The role of collagen in cancer: from bench to bedside. J Transl Med 17:309
2. Ferreira AM, Gentile P, Chiono V et al (2012) Collagen for bone tissue regeneration. Acta Biomater 8(9):3191–3200
3. Koohestani F, Braundmeier AG, Mahdian A et al (2013) Extracellular matrix collagen alters cell proliferation and cell cycle progression of human uterine leiomyoma smooth muscle cells. PLoS ONE 8(9):e75844
4. Mathew-Steiner SS, Roy S, Sen CK (2021) Collagen in wound healing. Bioengineering 8(5):63
5. Elsaid KA, Chichester CO (2006) Review: collagen markers in early arthritic diseases. Clin Chim Acta 365:68–77
6. Serrano AL, Muñoz-Cánoves P (2010) Regulation and dysregulation of fibrosis in skeletal muscle. Exp Cell Res 316(18):3050–3058
7. Tung JC, Barnes JM, Desai SR et al (2015) Tumor mechanics and metabolic dysfunction. Free Radical Biol Med 79:269–280
8. Wynn TA (2007) Common and unique mechanisms regulate fibrosis in various fibroproliferative diseases. J Clin Invest 117:524–529
9. Selman M, Montano M, Ramos C et al (1986) Concentration, biosynthesis and degradation of collagen in idiopathic pulmonary fibrosis. Thorax 41(5):355
10. Nikolov A, Popovski N (2022) Extracellular matrix in heart disease: Focus on circulating collagen type I and III derived peptides as biomarkers of myocardial fibrosis and their potential in the prognosis of heart failure: a concise review. Metabolites 12(4):297
11. Williams L, Layton T, Yang N et al (2022) Collagen VI as a driver and disease biomarker in human fibrosis. FEBS J 289(13):3603–3629
12. Gelse K (2003) Collagens—structure, function, and biosynthesis. Adv Drug Deliv Rev 55:1531–1546
13. Li Z, Bai X, Fan Y et al (2022) Structure of type II collagen from sturgeon cartilage and its effect on adjuvant-induced rheumatoid arthritis in rats. Food Funct 13(11):6152–6165
14. Xu R, Wu J, Zheng L et al (2023) Undenatured type II collagen and its role in improving osteoarthritis. Ageing Res Rev 91:102080
15. Prockop DJ, Kivirikko KI (1995) Collagens: molecular biology, diseases, and potentials for therapy. Annu Rev Biochem 64:403–434
16. Ezeani M, Noor A, Alt K et al (2021) Collagen-targeted peptides for molecular imaging of diffuse cardiac fibrosis. J Am Heart Assoc 10(18):e022139
17. Aronoff MR, Hiebert P, Hentzen NB et al (2021) Imaging and targeting LOX-mediated tissue remodeling with a reactive collagen peptide. Nat Chem Biol 17(8):865–871
18. Salarian M, Ibhagui OY, Yang JJ (2020) Molecular imaging of extracellular matrix proteins with targeted probes using magnetic resonance imaging. Wiley Interdisc Rev: Nanomed Nanobiotechnol 12(4):e1622
19. Srinivas GR, Barrach HJ, Chichester CO (1993) Quantitative immunoassays for type II collagen and its cyanogen bromide peptides. J Immunol Methods 159:53–62
20. Lin CY, Wang YL, Chen YJ et al (2022) Collagen-binding peptides for the enhanced imaging, lubrication and regeneration of osteoarthritic articular cartilage. Nat Biomed Eng 6(10):1105–1117

21. Xu K, Jin Y, Li Y et al (2022) Recent progress of exosome isolation and peptide recognition-guided strategies for exosome research. Front Chem 10:844124
22. Andrades JA, Santamaria JA, Wu LT et al (2001) Production of a recombinant human basic fibroblast growth factor with a collagen binding domain. Protoplasma 218:95–103
23. De Souza SJ, Brentani R (1992) Collagen binding site in collagenase can be determined using the concept of sense-antisense peptide interactions. J Biol Chem 267:13763–13767
24. Katsumata K, Ishihara J, Mansurov A et al (2019) Targeting inflammatory sites through collagen affinity enhances the therapeutic efficacy of anti-inflammatory antibodies. Sci Adv 5:eaay1971
25. Chan JM, Zhang L, Tong R et al (2010) Spatiotemporal controlled delivery of nanoparticles to injured vasculature. Proc Natl Acad Sci USA 107:2213–2218
26. Rothenfluh DA, Bermudez H, O'Neil CP, Hubbell JA (2008) Biofunctional polymer nanoparticles for intra-articular targeting and retention in cartilage. Nat Mater 7:248–254
27. Wei W, Li D, Cai X et al (2020) Highly specific recognition of denatured collagen by fluorescent peptide probes with the repetitive Gly-Pro-Pro and Gly-Hyp-Hyp sequences. J Mater Chem B 8:10093–10100
28. Wang AY, Mo X, Chen CS et al (2005) Facile modification of collagen directed by collagen mimetic peptides. J Am Chem Soc 127:4130–4131
29. Wei W, Li D, Cai X et al (2020) Peptide probes with aromatic residues Tyr and Phe at the X position show high specificity for targeting denatured collagen in tissues. ACS Omega 5:33075–33082
30. Li Y, Foss CA, Summerfield DD et al (2012) Targeting collagen strands by photo-triggered triple-helix hybridization. Proc Natl Acad Sci USA 109:14767–14772
31. Cai X, Liu Z, Zhao S et al (2017) A single stranded fluorescent peptide probe for targeting collagen in connective tissues. Chem Commun 53:11905–11908
32. Cai X, Wei W, Liu Z et al (2020) In situ imaging of pathological collagen by electrostatic repulsion-destabilized peptide probes. ACS Appl Bio Mater 3:7492–7499
33. Cai X, Wei W, Bi Y et al (2020) An intrinsically nontrimerizing peptide probe for specifically targeting pathological collagen in connective tissues. Adv Funct Materials 30:2004532
34. Kessler JL, Li Y, Fornetti J et al (2020) Enrichment of collagen fragments using dimeric collagen hybridizing peptide for urinary collagenomics. J Proteome Res 19:2926–2932
35. Takita KK, Fujii KK, Kadonosono T et al (2018) Cyclic peptides for efficient detection of collagen. ChemBioChem 19:1613–1617

Chapter 10
Raman Detection of Collagen Biomarkers

Abstract Raman scattering, a process where photons interact with molecules altering their energy levels, underpins Raman spectroscopy, a potent analytical technique for detecting analytes via molecular vibrations. This method has been instrumental in analyzing native collagen, including type I and type IV collagens, as well as studying the effect of temperature on type I collagen conformation. Surface-enhanced Raman spectroscopy (SERS) represents a significant advancement in molecular analysis, overcoming the inherent limitations of weak Raman signals. Recent progress in bioanalysis has been driven by robust SERS tags, incorporating essential elements such as plasmonic nanoparticle cores, Raman reporters, protective coatings, and targeting molecules. Dual SERS probes, such as Ag@R1@PCTP (S-P) and Ag@R2@ICTP (S-I), serve as precise biosensors for detecting pathological type I collagen in connective tissues. Additionally, stable SERS/fluorescent dual-mode peptide probes (SF-I, SF-IV, and SF-D) enable accurate detection and profiling of type I, type IV, and denatured collagen, facilitating hepatic fibrosis staging. The advancements in Raman spectroscopy and SERS have revolutionized biomolecule analysis, offering promising avenues for further breakthroughs in clinical diagnostics.

Keywords Raman spectroscopy · Surface-enhanced Raman spectroscopy · SERS tags · SERS probes · Biosensor

10.1 Introduction to Raman Spectroscopy

Raman scattering is an inelastic scattering process in which a photon interacts with a molecule, causing the photon to lose or gain energy. There are two types of Raman scattering: Stokes and anti-Stokes scattering (Fig. 10.1). In Stokes Raman scattering, the molecule gains energy from the incident photon, causing the scattered photons to lose energy and resulting in a shift to lower frequency or longer wavelength. Conversely, in anti-Stokes Raman scattering, the molecule loses energy as it relaxes to a lower vibrational level, transferring this energy to the scattered photon and causing a decrease in its wavelength. Stokes Raman scattering is more pronounced

© Springer Nature Singapore Pte Ltd. 2024

J. Xiao, *Collagen Mimetic Peptides and Their Biophysical Characterization*, Springer Series in Biomaterials Science and Engineering 20, https://doi.org/10.1007/978-981-10-1941-8_10

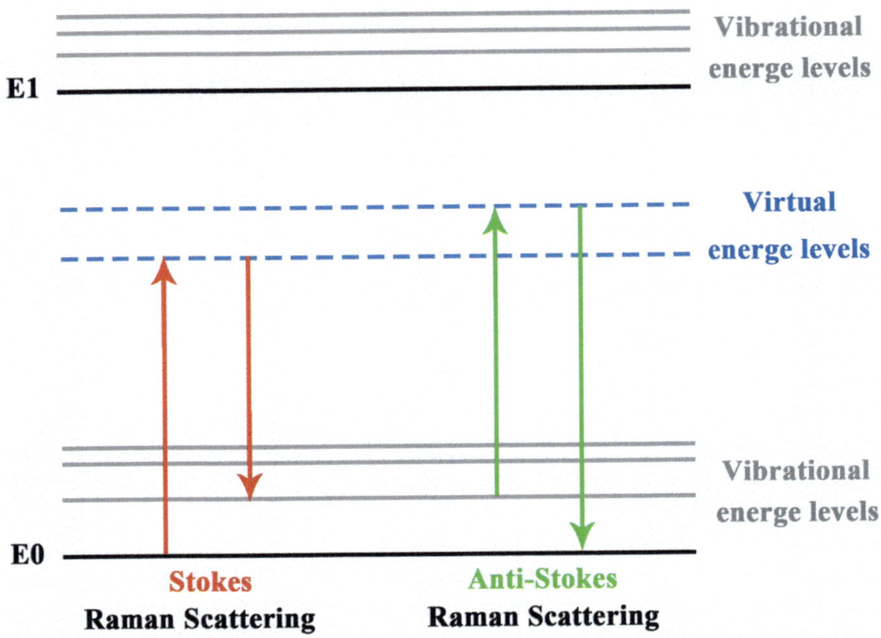

Fig. 10.1 Energetic transitions involved in Raman scattering

and easily detectable due to the higher number of molecules typically found in the vibrational ground state energy level compared to the vibrational excited state [1–3].

A standard Raman system consists of four main components: an excitation source, Rayleigh filter, microscope, and detector (Fig. 10.2). The choice of the excitation source depends on the sample's compatibility with higher photon energies, as well as the desired sensitivity and resolution for the study. Rayleigh filters are utilized to eliminate the more intense Rayleigh scattering, which can overshadow Raman scattering. A high-magnification objective provides superior axial spatial resolution suited for thin samples, whereas a low-magnification objective may be preferable for bulkier specimens. Charge-coupled devices (CCDs) serve as the primary detectors, gathering charge from scattered photons, with the magnitude of this charge directly correlating to the intensity of Raman scattering [4].

Raman spectroscopy is a robust analytical technique rooted in Raman scattering, facilitating the detection of analytes through their distinctive molecular bond vibrations [4, 5]. The peaks evident in the Raman spectrum correspond to specific molecular bonds, allowing for the identification and analysis of the chemical composition in a given sample, while their magnitude directly correlates with the molecule concentration, enabling quantitative assessment. Moreover, the Raman spectrum of proteins consists of discrete bands corresponding to vibrational modes of the peptide backbone and its side chains, providing unique insights into the conformational elements that define protein secondary structure [6]. Raman investigations into protein structures focus on establishing correlations between the positions of amide I and amide

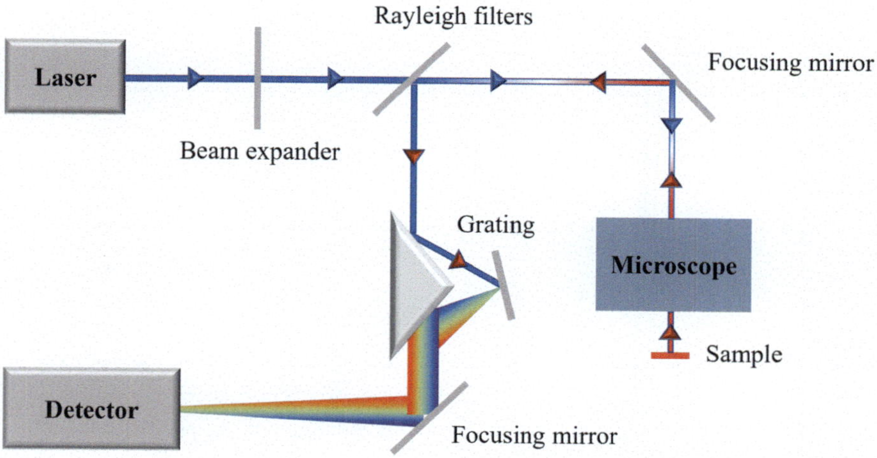

Fig. 10.2 The composition of confocal Raman microscope: laser; Rayleigh fibers; microscope; CCD detector

III vibrations and the proportions of each secondary structural component. Notably, characteristic wavenumbers emerge: 1662–1655 and 1272–1264 cm^{-1} for α-helix and 1674–1672 and 1242–1227 cm^{-1} for β-sheet structures, respectively, in both amide I and amide III modes [7]. The non-destructive and label-free nature of Raman spectroscopy has solidified its role as an essential tool in protein research, providing a wealth of information crucial for deciphering the intricate molecular mechanisms underlying protein folding, unfolding, and interactions [8].

10.2 Raman Characterization of Collagen Conformation

Raman spectroscopy has been employed to analyze native collagen, revealing three distinct regions within its spectrum: lipids, amides, and proline-hydroxyprolin (Fig. 10.3) [9]. The recorded peaks at 2881 and 2933 cm^{-1} within the lipid region correspond to CH_2 stretching and C–H vibration, respectively, whereas the proline and hydroxyproline region reveals distinct side-chain peaks at 925 and 860 cm^{-1} for proline, as well as 880 cm^{-1} for hydroxyproline. Notably, the amide region displays two peaks at 1642 and 1630 cm^{-1} for Amide I vibrations, along with a peak at 1271 cm^{-1} for Amide III vibrations, showcasing a hallmark of the triple helical structure. Raman spectroscopy presents a robust approach to characterizing the distinct amino acids and triple helical conformation inherent in native collagen.

Comparative Raman analysis of type I and type IV collagens unveils striking disparities in their spectra (Fig. 10.4) [10]. Remarkable distinctions emerge in the wavenumbers of amide I peak, signifying variances in amino acid proportions at the X and Y positions. Additionally, the relative intensity ratio of the amide III vibration

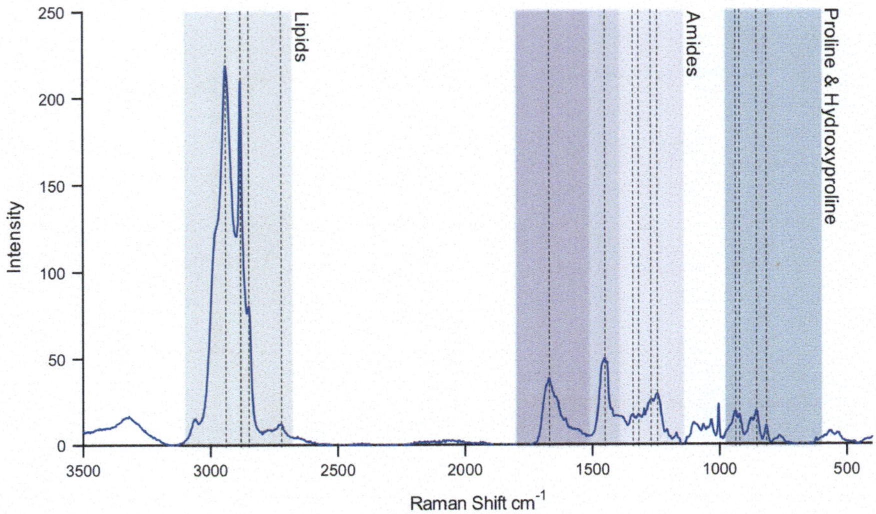

Fig. 10.3 Raman spectrum of native collagen [9]. Reprinted from ref. [9], Copyright © 2019, with permission from Taylor & Francis Group

peaks (I_{1271}/I_{1246}) exhibits pronounced differences, suggesting significant variations in the proline content between the collagen types. Notably, collagen IV demonstrates heightened intensity peaks at 1004 and 1033 cm^{-1}, characteristic of phenylalanine, implying a potentially elevated phenylalanine composition. Moreover, the presence of peaks at 510 and 540 cm^{-1}, ascribed to S–S vibrations, exclusively in type IV collagen, confirms the existence of disulfide bridges within its 7S domains. These findings underscore a marked divergence in amino acid distribution, encompassing proline, phenylalanine, and cysteine, between the two distinct collagen types.

Dong and colleagues conducted a comprehensive analysis to investigate the impact of temperature on the conformation of type I collagen, covering a broad range from −150 to 200 °C (Fig. 10.5) [11]. The stability of the band at 1003 cm^{-1}, attributed to phenylalanine, remains remarkably consistent throughout various temperature conditions, displaying minimal sensitivity to temperature changes. With increasing temperature, the bands at 1655, 1442, 1123, 1060, 917, and 814 cm^{-1} all shift to lower wavenumbers, while the band at 1302 cm^{-1}, related to CH_2 deformation, shifts to higher wavenumbers. Remarkably, the 1263 cm^{-1} band indicative of Amide III vibration vanishes when the temperature exceeds 110 °C, signaling the disruption of hydrogen bonds within collagen's triple helices and a subsequent increase in protein structural disorder.

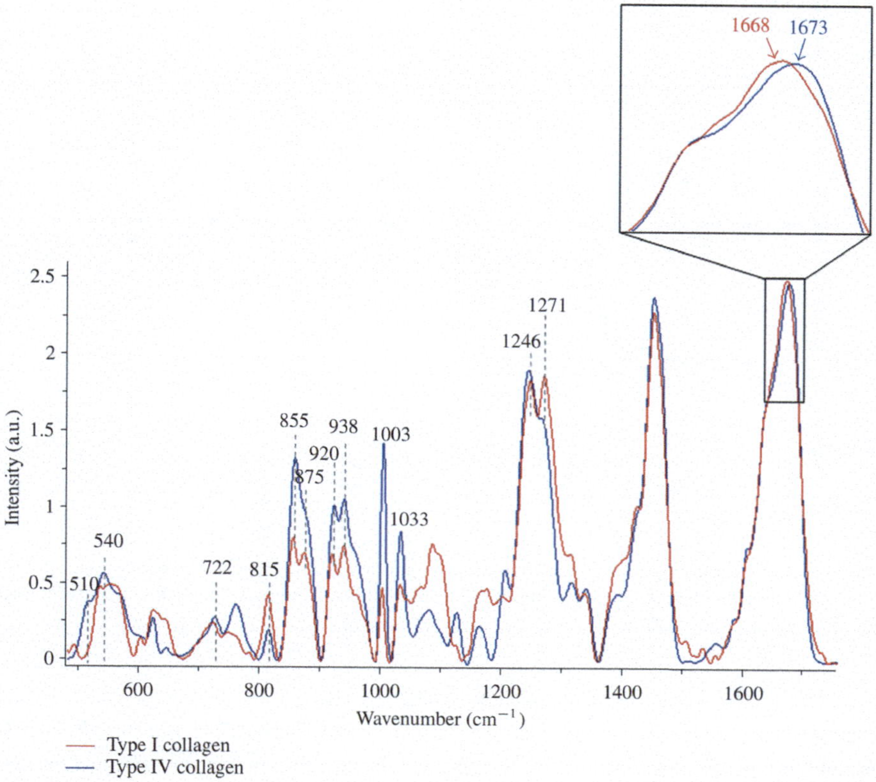

Fig. 10.4 Raman spectra of type I collagen (red) and IV collagen (blue) [10]. Reprinted from ref. [10], Copyright © 2012, with permission from Hindawi Ltd.

10.3 Introduction to SERS

Surface-enhanced Raman spectroscopy (SERS) represents a transformative breakthrough in molecular analysis, born out of the need to surmount the intrinsic limitations of the Raman effect. In conventional Raman spectroscopy, a mere one in 10^7 incident photons undergo Raman scattering, resulting in feeble Raman signals, particularly in situations involving low analyte concentrations. The presence of intense fluorescence from molecules further compounds this challenge, often masking the desired Raman signals. However, SERS emerged as an ingenious remedy to these challenges, with its origins dating back to 1974 when Fleischmann and colleagues serendipitously discovered an astonishing amplification of Raman signals emanating from pyridine molecules in close proximity to a roughened silver electrode [12]. These pioneering endeavors collectively ushered in the era of SERS, a revolutionary technique capable of magnifying Raman signals by orders of magnitude.

SERS finds its foundation in the utilization of metallic colloidal nanoparticles or roughened metal surfaces, which serve as substrates for the adherence of molecules

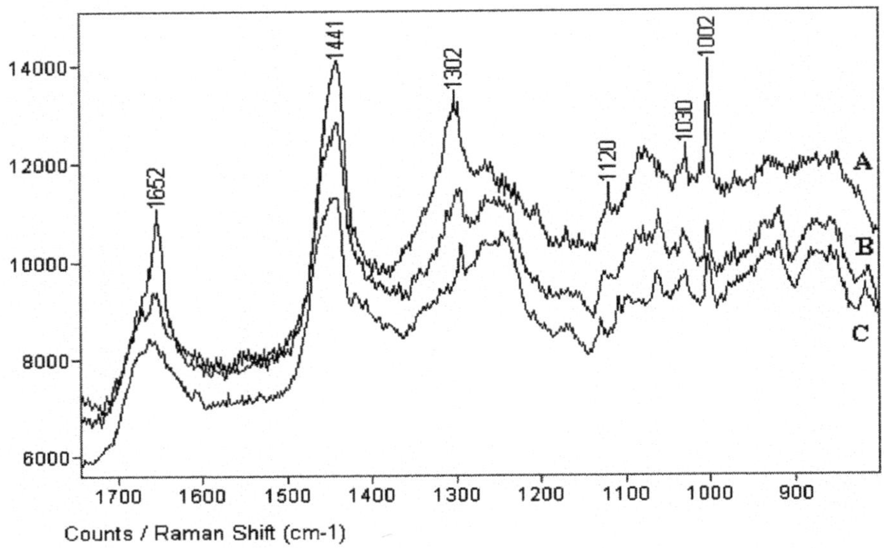

Fig. 10.5 Raman fingerprint region of type I collagen exposed to different temperatures: 80 °C (**A**), 20 °C (**B**) and −100 °C (**C**) [11]. Reprinted from ref. [11], Copyright © 2004, with permission from Elsevier Science Ltd.

(Fig. 10.6). The exceptional sensitivity of SERS can be attributed to two central enhancement mechanisms: the electromagnetic mechanism (EM) and the chemical mechanism (CM) [13, 14]. The EM mechanism emerges from the intricate interplay between incident light and plasmon resonances within metallic nanoparticles, resulting in the creation of highly concentrated local electromagnetic fields near the particle surfaces. This leads to a remarkable enhancement of Raman signals, often reaching the extraordinary range of 10^{10}–10^{11} times. In contrast, the CM mechanism arises from chemical interactions between plasmonic materials, frequently in the form of metallic nanoparticles, and the molecules of interest. While the CM mechanism exerts less dominance in the overall enhancement compared to its electromagnetic counterpart, it assumes a pivotal role in specific SERS applications, offering an enhancement factor of approximately 10^3.

SERS seamlessly integrates the intrinsic advantages of Raman spectroscopy, including its capacity to identify molecules through their distinctive vibrational signatures, non-destructive characteristics, minimal sample preparation prerequisites, suitability for measurements in aqueous environments due to the weak water spectrum, ability to concurrently detect a multitude of analytes, and the potential for on-site analysis using portable instrumentation. By combining the information-rich nature of Raman spectra with the signal amplification provided by plasmonic nanoparticles, SERS has evolved into a potent analytical technique, occasionally achieving single-molecule detection, with diverse applications in chemistry, biology, materials science, and environmental monitoring [15, 16].

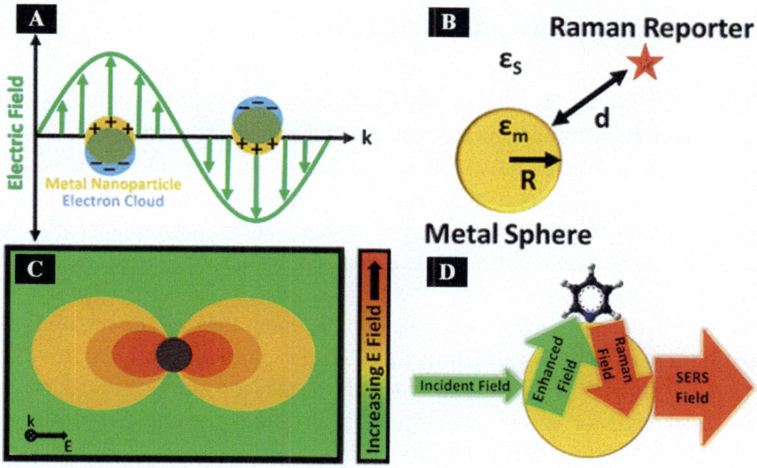

Fig. 10.6 **A** Illustration of the oscillating electron cloud. **B** The significant dependence of the EM enhancement on the particle size. **C** Emitted dipole field of a metallic nanoparticle under light excitation. **D** EM enhancement of both the incident field and the scattered field [16]. Reprinted from ref. [16], Copyright © 2015, with permission from American Chemical Society

10.4 Design of SERS Probes

The application of the SERS technique in the field of biomedicine encounters substantial challenges, primarily stemming from the remarkably low intrinsic Raman cross-sections of biomolecules. Recent advancements in bioanalysis been propelled by the innovation of resilient SERS tags, which entail critical elements including the selection of the plasmonic nanoparticle core, the choice of Raman reporters, the design of a protective coating shell, and the integration of targeting molecules (Fig. 10.7) [17]. The plasmonic nanoparticle core plays a central role in enhancing Raman signals, with its chemical composition, size, and shape significantly influencing the performance of SERS tags and enabling the quantification of analytes through amplified Raman signals from the reporters. The protective coating shell serves to prevent coalescence and fouling, ensuring stable SERS signals, while targeting molecules on the surface of the plasmonic nanoparticle core confer specific functionality to the SERS tags.

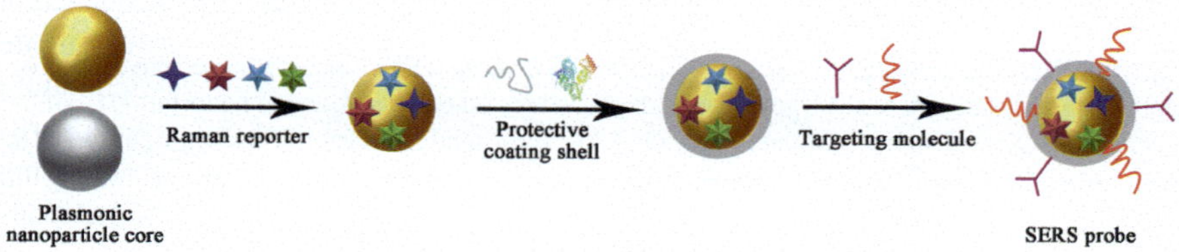

Fig. 10.7 Schematic illustration of the general design of SERS probes

10.4.1 Plasmonic Nanoparticle Cores

Gold and silver nanoparticles are widely favored as plasmonic nanoparticle cores in SERS tags due to their robust electromagnetic enhancement capabilities. Moreover, gold and silver nanospheres have gained significant attention in both research and practical applications, thanks to their easily controlled size distribution, prolonged stability, and outstanding biocompatibility. The size of these nanospheres plays a critical role in influencing their SERS signal enhancement capacity, with larger volumes enhancing the electromagnetic field; however, excessive size leads to greater radiation damping effects and a decrease in the enhancement factor, with the optimal size typically falling within the 30–100 nm range. While spherical nanoparticles exhibit a limited range of adjustable localized surface plasmon resonance (LSPR) bands, particles with sharp edges or vertices offer greater versatility, spanning from the visible to Near-Infrared (NIR) regions. Distinctive shapes like nanorods, nanocubes, nanostars, nanorice, nanotriangle plates, and nanoflowers can achieve remarkable SERS enhancements, reaching values as high as 10^{11} (Fig. 10.8) [18–21]. Generally, silver outperforms gold in enhancing Raman signals, producing SERS signals that are 10 to 100 times more intense than those generated by similar gold nanostructures, owing to its UV-region d-s band gap, which minimizes plasmon mode damping. Conversely, gold nanoparticles are often the preferred choice for in-vivo applications due to their low cytotoxicity, high biocompatibility, and ease of functionalization [22–24].

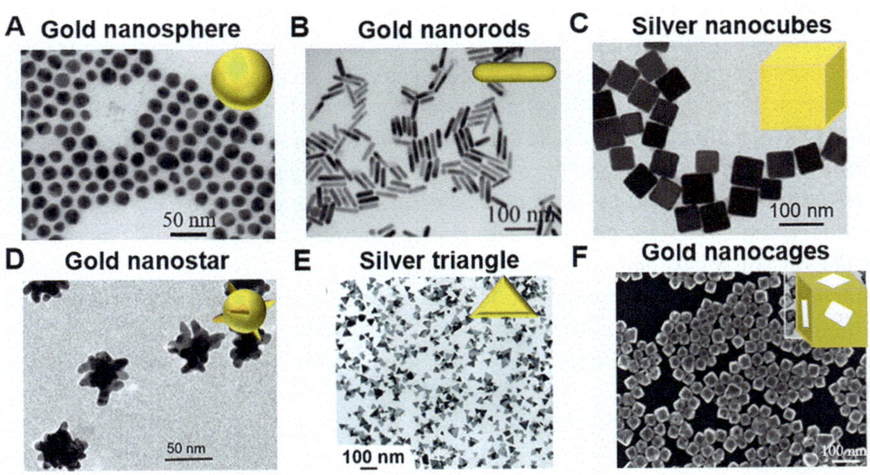

Fig. 10.8 Representative colloidal plasmonic nanostructures. TEM images and tunable extinction spectra of gold nanospheres (**A**), gold nanorods (**B**), silver nanocubes (**C**), gold nanostars (**D**), silver triangle nanoplate (**E**) and gold nanocages (**F**) [25]. Reprinted from ref. [25], Copyright © 2018, with permission from Elsevier Science Ltd.

10.4.2 Raman Reporters

Raman reporter molecules are pivotal in creating the unique spectral signature of SERS tags, enabling the indirect identification of specific targets. These reporters possess specific attributes to ensure their efficacy, including (1) a large Raman scattering cross-section for generating intense signals, (2) a limited number of Raman peaks to minimize spectral overlap in multiplexing experiments, (3) high stability when exposed to laser irradiation, and (4) the ability to bind to the surface of plasmonic nanoparticles through chemical bonds or physical interactions. Typically, aromatic molecules containing nitrogen or sulfur elements are preferred Raman reporters, facilitating the construction of a diverse Raman reporter library with various Raman codes for in vitro bioanalysis. However, when crafting SERS tags for in vivo applications, careful consideration must be given to potential Raman shifts from endogenous biological components in the fingerprint region, with the aim of preventing unwanted spectral overlap with SERS tag frequencies. To mitigate this issue, a specialized class of bioorthogonal Raman reporters has been developed, excelling in generating strong Raman signals in biologically silent regions. Additionally, alkyne-based reporters, when linked to various organic structures, exhibit distinct Raman shifts within the silent region, significantly expanding the range of distinguishable SERS tags through alkyne encoding, and they are widely utilized as Raman reporters for visualizing bioactive molecules [17].

10.4.3 Protective Coating Shell

A protective coating shell is essential to safeguard Raman reporters on nanoparticle substrates in complex biochemical environments, addressing challenges like reporter dissociation and interference from environmental molecules, thereby ensuring the stability and reliability of SERS tags for advanced applications (Fig. 10.9) [21, 26–28]. Various materials, including silica, polymers, biomolecules, and liposomes, have been employed for surface coating to enhance the stability and biocompatibility of SERS tags. Silica coating, initially developed and widely commercialized as an encapsulation method for SERS tags, offers advantages such as high stability, good water solubility, minimal nonspecific binding, and ease of modification in biological systems [29, 30]. Polymer coatings, especially SH-PEG, are commonly used in SERS tags due to their biocompatibility, stability, and accessibility, along with benefits like adjustable coating thickness, low affinity for interfering molecules, excellent in vivo biodistribution, and the potential for surface functionalization through chemical end groups [31]. Biomolecules, such as BSA (bovine serum albumin), serve as coating shells to enhance the biocompatibility of SERS tags by adhering to metal surfaces through weak interactions and forming protective layers around the nanoparticles [32]. Liposomes represent a promising choice for particle coating due to their intrinsic biocompatibility, self-assembly capability into structured arrangements, and

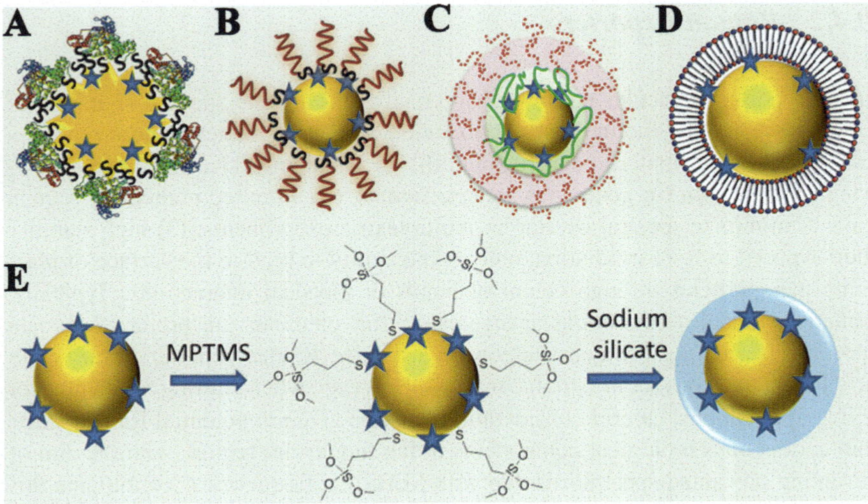

Fig. 10.9 Strategies for surface-coating SERS probes include coating with denatured BSA (**A**), SH-PEG (**B**), amphiphilic diblock copolymer (**C**), liposome (**D**) and silica shells (**E**) [28]. Reprinted from ref. [28], Copyright © 2013, with permission from American Chemical Society

potential to provide stability and targeting features through both the lipid vesicle and attached targeting molecules [33–35].

10.4.4 Targeting Molecules

Functionalizing SERS tags with targeting molecules like antibodies, aptamers, peptides, and small-molecule ligands is crucial for enabling the tags to exclusively recognize the intended target and eliminating nonspecific binding [36, 37]. To prevent the exchange of targeting moieties between different tags, secure binding, often facilitated by covalent bonds with capping layer molecules, is essential. Antibodies are frequently employed in SERS tags due to their ability to target a wide range of substances, and they can be easily attached to plasmonic nanoparticles through the reaction between carboxylic acid groups and amine groups, using reagents like 1-ethyl-3-(3-dimethyl aminopropyl) carbodiimide and N-hydroxy succinimides. Aptamers, known for their durability compared to antibodies, can be affixed to nanoparticles tagged with streptavidin through the strong biotin-streptavidin interaction. Additionally, Cysteine-terminated peptides can be conjugated to the BSA layer through the formation of S–S bonds [38–40].

10.5 SERS Detection of Collagen

10.5.1 SERS Probes for Targeting Pathological Type I Collagen

Our research group has introduced dual SERS probes, Ag@R1@PCTP (S-P) and Ag@R2@ICTP (S-I), as robust biosensors for accurately identifying pathological type I collagen in connective tissues (Fig. 10.10). These SERS probes were created by combining Ag NPs as the plasmonic core with R1 (S-(4-((trimethylsilyl)ethynyl)phenyl) ethanethioate) and R2 (S-(4-cyanophenyl) ethanethioate) as Raman reporters, along with targeting peptide sequences, PCTP (Cys-Ahx-$(GPO)_7$G) for pathological collagen and ICTP (Cys-Ahx-LRELHLNNNG) for type I collagen. Raman reporters, alkyne R1 and nitrile R2, were linked to Ag NPs through sulfur bonds, enabling them to display distinctive narrow bands in the Raman silent region (1800–2800 cm^{-1}) and thereby facilitating precise SERS detection (Fig. 10.10). Binding studies revealed that Ag@R1@PCTP possessed impressive targeting capability specifically tailored for pathological collagen, whereas Ag@R2@ICTP effectively recognized both intact and pathological type I collagen.

Raman imaging of tissues co-stained with Ag@R1@PCTP and Ag@R2@ICTP vividly illustrated pathological collagen distribution in the blue channel and type I collagen distribution in the green channel, while merged images unequivocally confirmed the presence of pathological type I collagen through overlapping cyan signals (Fig. 10.10). In normal tail tissues co-stained with Ag@R1@PCTP and Ag@R2@ICTP, strong signals in the green channel validated the presence of type I collagen, with a faint 2154 cm^{-1} peak at points 1 and 2 indicating the absence of pathological collagen. Conversely, Raman micrographs of impaired tail tissue co-stained with Ag@R1@PCTP and Ag@R2@ICTP exhibited prominent signals in the cyan channel, emphasizing the precise targeting of pathological type

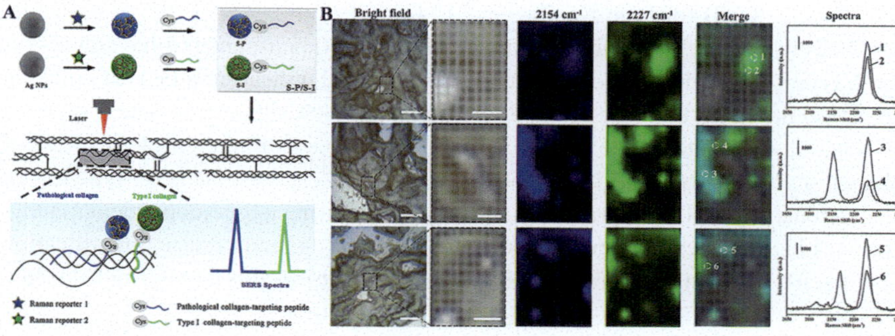

Fig. 10.10 **A** Construction of SERS probes and precise imaging of pathological type I collagen. **B** SERS imaging of probe S-P (Ag@R1@PCTP) and S-I (Ag@R2@ICTP) simultaneously

I collagen using dual SERS probes. Furthermore, Raman micrographs of impaired tendon tissue co-stained with Ag@R1@PCTP and Ag@R2@ICTP also displayed cyan channel signals, confirming the accurate localization of pathological type I collagen with dual SERS probes in impaired tendon tissues. The simultaneous use of Ag@R1@PCTP and Ag@R2@ICTP in tissue staining assays generated dual-channel signals, enabling precise imaging of pathological type I collagen and holding significant promise for diagnosing collagen I-related diseases.

10.5.2 Fluorescence-Guided SERS Imaging for Multiplex Collagen

Our research group has created stable SERS/fluorescent dual-mode peptide probes (SF-I, SF-IV, and SF-D) for the precise detection and profiling of type I, type IV, and denatured collagen, facilitating accurate hepatic fibrosis staging. These probes consist of Ag NPs serving as SERS substrates, Raman reporters, and FAM-labeled collagen-targeting peptides (CTP), featuring multiple charged Asp's at the N-terminal, antibody mimetic peptide sequences, an Ahx (aminohexanoic acid) linker, and Cys at the C-terminal. Antibody mimetic peptides have shown selective recognition for various collagen types, including type I, type IV, and denatured collagen, offering advantages such as easy synthesis, modification, and strong penetration. The inclusion of Asp′s is utilized to maintain the negative charge state of CTP, resulting in effective electrostatic repulsion that enhances the dispersibility and stability of SERS probes. Through the conjugation of FAM-labeled peptide probes and Raman reporters to the surface of Ag NPs via Ag–S bonds, SERS/fluorescent dual-mode probes, specifically targeting type I, type IV, and denatured collagen, were successfully developed.

The SF-I, SF-IV, and SF-D probes were harnessed to stain liver fibrotic tissues across S0-S4 stages, facilitating fluorescence-guided SERS imaging for multiple collagen types (Fig. 10.11). These probes illuminated tissues by capturing signals at 2227 cm^{-1} (red) for type I collagen, 2154 cm^{-1} (yellow) for type IV collagen, and 2102 cm^{-1} (blue) for denatured collagen. In the S0 phase, negligible signals for type I, type IV, and denatured collagen were detected, indicating a lack of collagen deposition in the portal area. In the S1 phase, significant signals for type I, type IV, and denatured collagen were observed, indicating the accumulation of all forms of collagen in fibrotic tissues. Throughout the S2-S4 phases, the signals for type I, type IV, and denatured collagen gradually increased, exhibiting a type-dependent manner (Fig. 10.11). The specific identification, spatial mapping, and precise quantification of collagen types using the three probes offered valuable insights into the multifaceted roles played by various collagen subtypes in liver fibrosis.

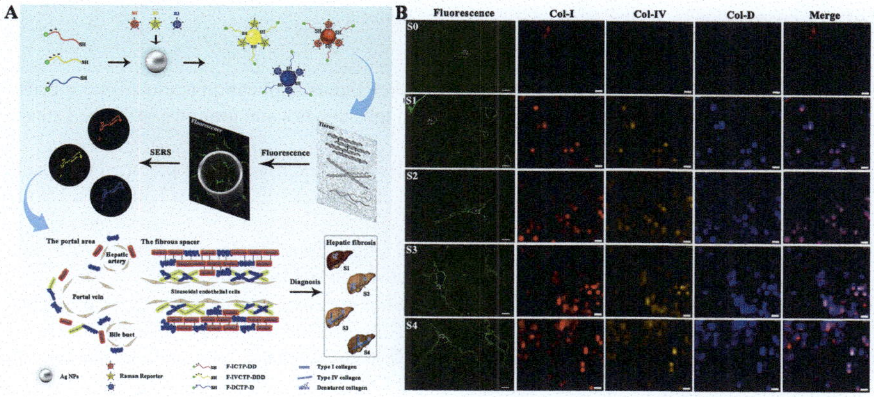

Fig. 10.11 **A** Construction of SERS/fluorescent dual-mode peptide probes SF-I, SF-IV, and SF-D. **B** Fluorescence-guided SERS imaging of S0-S4 hepatic fibrosis

10.6 Protocol for SERS Imaging of Collagen in Tissues

- **Preparation of SERS probe Solutions**: Prepare solutions of SERS probe peptides with a concentration of 15 μM using deionized water. Store the solutions in the dark at 4 °C until ready to use.
- **Tissue Preparation**: Incubate the tissue slides in 1 × PBS buffer for 5 min at room temperature.
- **Blocking**: Place the tissue slides in a humidified chamber. Apply 0.5 mL of blocking solution to each slide. Incubate the slides for 30 min at room temperature. Remove the blocking solution by blotting the slides on a paper towel.
- **Staining**: Apply 100 μL of probe peptide solution to cover the tissue sections of each slide. Incubate the slides for 2 min to allow the probe peptide to permeate the tissues. Cover the tissue sections of each slide with a piece of parafilm to prevent drying. Place the slides in a humidified chamber and incubate them at 4 °C for 4 h.
- **Removal of Excess Staining Solution**: Gently remove the parafilm with forceps. Blot the slides on a paper towel to remove excess S-D solution.
- **Washing**: Immerse the slides in 1 × PBS buffer in a staining jar for 5 min. Repeat this step three times in fresh 1 × PBS to wash off unbound staining agents.
- **Mounting**: Add a drop of mounting medium to each tissue section. Cover with a glass cover slip, avoiding air bubbles.
- **Imaging**: Image the tissue slides using a SERS microscope.

References

1. Browne WR, McGarvey JJ (2007) The Raman effect and its application to electronic spectroscopies in metal-centered species: Techniques and investigations in ground and excited states. Coord Chem Rev 251:454–473
2. Holze R , Smith E, Dent G (eds) (2005) Modern Raman spectroscopy—a practical approach, Wiley, Chichester, United Kingdom, 2005, 210 + XI p., 24.95 £; ISBN 0471497940. J Solid State Electrochem 11:558–558
3. Clark RJ (2013) Rayleigh, Ramsay, Rutherford and Raman–their connections with, and contributions to, the discovery of the Raman effect. Analyst 138:729–734
4. Butler HJ, Ashton L, Bird B et al (2016) Using Raman spectroscopy to characterize biological materials. Nat Protoc 11:664–687
5. Khmaladze A, Ganguly A, Kuo S et al (2013) Tissue-engineered constructs of human oral mucosa examined by Raman spectroscopy. Tissue Eng Part C Methods 19:299–306
6. Schlucker S (2014) Surface-enhanced Raman spectroscopy: concepts and chemical applications. Angew Chem Int Ed 53:4756–4795
7. McColl IH, Blanch EW, Gill AC, Rhie AG, Ritchie MA, Hecht L, Nielsen K, Barron LD (2003) A new perspective on β-sheet structures using vibrational Raman optical activity: from poly(l-lysine) to the prion protein. J Am Chem Soc 125:10019–10026
8. Vlasov AV, Maliar NL, Bazhenov SV et al (2020) Raman scattering: from structural biology to medical applications. Crystals 10(1):38
9. Martinez MG, Bullock AJ, MacNeil S et al (2019) Characterisation of structural changes in collagen with Raman spectroscopy. Appl Spectrosc Rev 54:509–542
10. Nguyen TT, Gobinet C, Feru J et al (2012) Characterization of type I and IV collagens by Raman microspectroscopy: identification of spectral markers of the dermo-epidermal junction. Spectro Int J 27:421–427
11. Dong R, Yan X, Pang X, Liu S (2004) Temperature-dependent Raman spectra of collagen and DNA. Spectrochim. Acta A Mol Biomol Spectrosc 60:557–561
12. M. Fleischmann P J H, A.J. McQuillan, (1974) Raman spectra of pyridine adsorbed at a silver electrode. Chem Phys Lett 26:163–166
13. Willets KA, Van Duyne RP (2007) Localized surface plasmon resonance spectroscopy and sensing. Annu Rev Phys Chem 58:267–297
14. Sharma B, Frontiera RR, Henry A-I et al (2012) SERS: materials, applications, and the future. Mater Today 15:16–25
15. Cialla-May D, Zheng XS, Weber K et al (2017) Recent progress in surface-enhanced Raman spectroscopy for biological and biomedical applications: from cells to clinics. Chem Soc Rev 46:3945–3961
16. Lane LA, Qian X, Nie S (2015) SERS nanoparticles in medicine: from label-free detection to spectroscopic tagging. Chem Rev 115:10489–10529
17. Liu H, Gao X, Xu C et al (2022) SERS tags for biomedical detection and bioimaging. Theranostics 12:1870–1903
18. Zhang Q, Li N, Goebl J, Lu Z et al (2011) A systematic study of the synthesis of silver nanoplates: is citrate a "magic" reagent? J Am Chem Soc 133:18931–18939
19. Bao H, Zhang Q, Xu H et al (2016) Effects of nanoparticle size on antitumor activity of 10-hydroxycamptothecin-conjugated gold nanoparticles: in vitro and in vivo studies. Int J Nanomedicine 11:929–940
20. Qiang Zhang WL, Moran C, Zeng J et al (2010) Seed-mediated synthesis of Ag nanocubes with controllable edge lengths in the range of 30–200 nm and comparison of their optical properties. J Am Chem Soc 132:11372–11378
21. Li M, Cushing SK, Zhang J et al (2012) Shape-dependent surface-enhanced Raman scattering in gold-Raman probe-silica sandwiched nanoparticles for biocompatible applications. Nanotechnology 23:115501
22. Anselmo AC, Mitragotri S (2016) Nanoparticles in the clinic. Bioeng Transl Med 1:10–29

23. Kennedy DC, Orts-Gil G, Lai C-H, ller LM et al (2014) Carbohydrate functionalization of silver nanoparticles modulates cytotoxicity and cellular uptake. J Nanobiotechnol 12:1–8
24. Alkilany AM, Murphy CJ (2010) Toxicity and cellular uptake of gold nanoparticles: what we have learned so far? J Nanopart Res 12:2313–2333
25. Shan B, Pu Y, Chen Y et al (2018) Novel SERS labels: rational design, functional integration and biomedical applications. Coord Chem Rev 371:11–37
26. Skrabalak SE, Au L, Li X, Xia Y (2007) Facile synthesis of Ag nanocubes and Au nanocages. Nat Protoc 2:2182–2190
27. Huang X, Neretina S, El-Sayed MA (2009) Gold nanorods: from synthesis and properties to biological and biomedical applications. Adv Mater 21:4880–4910
28. Wang Y, Yan B, Chen L (2013) SERS tags: novel optical nanoprobes for bioanalysis. Chem Rev 113:1391–1428
29. Yin Y, Mei R, Wang Y et al (2020) Silica-coated, waxberry-like surface-enhanced Raman resonant scattering tag-pair with near-infrared Raman dye encoding: toward in vivo duplexing detection. Anal Chem 92:14814–14821
30. Zhang Y, Qiu Y, Lin L et al (2017) Ultraphotostable mesoporous silica-coated gap-enhanced Raman tags (GERTs) for high-speed bioimaging. ACS Appl Mater Interfaces 9:3995–4005
31. Choi N, Dang H, Das A et al (2020) SERS biosensors for ultrasensitive detection of multiple biomarkers expressed in cancer cells. Biosens Bioelectron 164:112326
32. Holler RPM, Jahn IJ, Cialla-May D et al (2020) Biomacromolecular-assembled nanoclusters: key aspects for robust colloidal SERS sensing. ACS Appl Mater Interfaces 12:57302–57313
33. Mei R, Wang Y, Liu W et al (2018) Lipid bilayer-enabled synthesis of waxberry-like core-fluidic satellite nanoparticles: toward ultrasensitive surface-enhanced Raman scattering tags for bioimaging. ACS Appl Mater Interfaces 10:23605–23616
34. Su X, Wang Y, Wang W et al (2016) Phospholipid encapsulated AuNR@Ag/Au Nanosphere SERS tags with environmental stimulus responsive signal property. ACS Appl Mater Interfaces 8:10201–10211
35. Stewart AF, Lee A, Ahmed A, Shell Ip et al. (2014) Rational design for the controlled aggregation of gold nanorods via phospholipid encapsulation for enhanced Raman scattering. ACS NANO 8:5462–5467
36. Attarwala H (2010) Role of antibodies in cancer targeting. J Nat Sci Biol Med 1:53–56
37. Mousavizadeh A, Jabbari A, Akrami M et al (2017) Cell targeting peptides as smart ligands for targeting of therapeutic or diagnostic agents: a systematic review. Colloids Surf B 158:507–517
38. Pal S, Harmsen S, Oseledchyk A et al (2017) MUC1 aptamer targeted SERS nanoprobes. Adv Funct Mater 27:1606632
39. Zheng XS, Zong C, Wang X et al (2019) Cell-penetrating peptide conjugated SERS nanosensor for in situ intracellular pH imaging of single living cells during cell cycle. Anal Chem 91:8383–8389
40. Wei C, Sun R, Jiang Y et al (2021) Protease-protection strategy combined with the SERS tags for detection of O-GlcNAc transferase activity. Sens Actuators B Chem 345:130410

Chapter 11
Heterotrimeric Peptide Models of Collagen

Abstract Deciphering the intricate structural and functional characteristics of heterotrimeric collagens relies on the development of heterotrimeric triple-helical peptide systems, yet incorporating three chains with distinct amino acid sequences remains challenging. This chapter offers a comprehensive exploration of heterotrimeric peptide models of collagen, covering their design, synthesis, and characterization methodologies. Various strategies, including Lys-Lys and Cys-Cys disulfide linkages, have been devised for covalently tethering heterotrimeric peptides. Additionally, it delves into the concept of charge-paired heterotrimeric peptides, detailing both AAB and ABC configurations and their implications for structural integrity and functionality. Computational simulation techniques are surveyed for refining heterotrimeric peptide design by precisely manipulating composition and register through optimization of pairwise interactions. Innovative Triblock peptides with three distinct domains are introduced, facilitating the creation of AAB-type and ABC-type heterotrimers with enhanced stability and greater flexibility in amino acid composition and arrangement. Lastly, the chapter explores covalent capture strategies for heterotrimeric peptides and provides insightful reflections on the current landscape and future trajectories of research in this dynamic field.

Keywords Heterotrimeric peptides · Covalently tethering · Charge-paired interaction · Computational optimization · Triblock heterotrimeric peptides

11.1 Introduction to Heterotrimeric Peptides

The construction and characterization of hybrid triple-helical peptide systems are essential for modeling heterotrimeric collagens and diseases involving one or two mutations in a collagen triple-helix [1, 2]. However, designing triple-helical peptides with three chains featuring different amino acid sequences to mimic heterotrimeric collagens, like Type I collagen in bone and Type IV collagen in basement membranes, poses significant challenges [3–6]. Early breakthroughs involved covalent synthetic methods such as Lys-Lys covalent tethering and disulfide bridging [7–10]. The

© Springer Nature Singapore Pte Ltd. 2024

J. Xiao, *Collagen Mimetic Peptides and Their Biophysical Characterization*,
Springer Series in Biomaterials Science and Engineering 20,
https://doi.org/10.1007/978-981-10-1941-8_11

first heterotrimeric collagen mimetic peptide, utilizing Lys-Lys covalent tethering, emerged in the mid-1990s, nearly 30 years after their homotrimeric counterparts [11]. More recently, noncovalent self-assembled peptides have become a hotspot in constructing collagen heterotrimers due to their advantages, including easy synthesis, high purity, and cost-effectiveness.

The creation of self-assembled heterotrimers is a complex task, surpassing the intricacy of homotrimers due to the need for precise control over both composition and register (Fig. 11.1). In a binary mixture of peptides A and B, four different species can emerge (A2B, AB2, A3, and B3). Complexity further increases with the addition of peptides A, B, and C, resulting in ten distinct triple-helical compositions (A3, B3, C3, A2B, AB2, A2C, AC2, B2C, C2B, and ABC). The introduction of one-residue staggering results in the coexistence of heterotrimers with different registers. For example, the A2B heterotrimer can assemble as BAA, ABA, or AAB, depending on the positioning of chain B. In a ternary mixture, a total of 27 unique helices can be formed, including six distinct registers of the ABC heterotrimer. Recent advances, such as the utilization of charged pairs, computational optimization, and triblock peptides, have emerged as exciting strategies to facilitate the self-assembly of peptides into heterotrimers.

Heterotrimer peptides have undergone thorough characterization regarding molecular structure, stability, folding, and dynamics using advanced biophysical

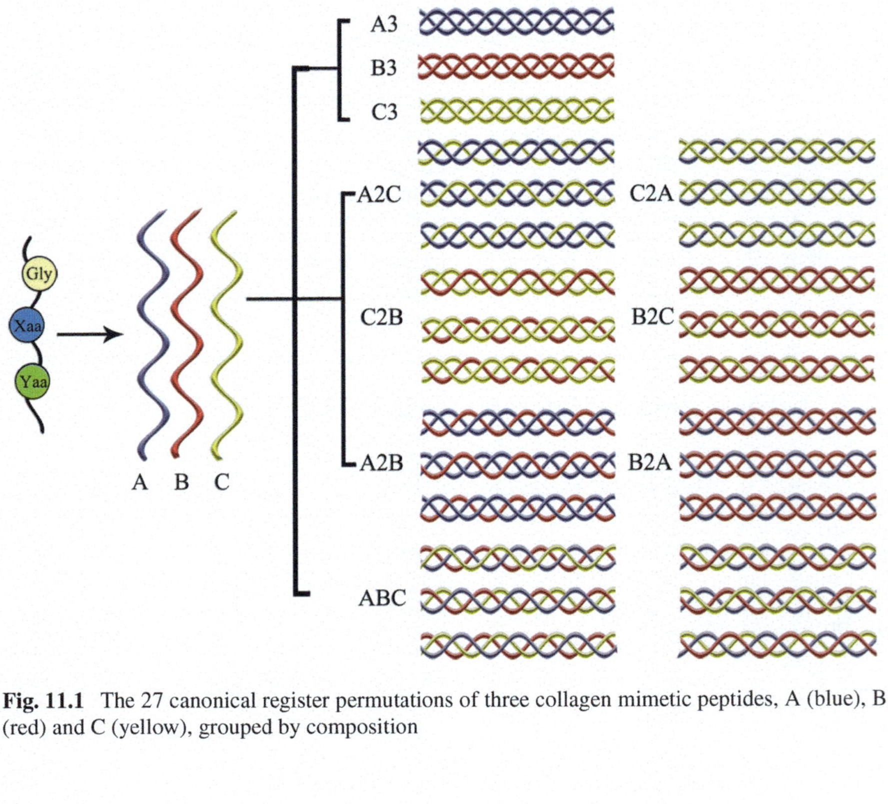

Fig. 11.1 The 27 canonical register permutations of three collagen mimetic peptides, A (blue), B (red) and C (yellow), grouped by composition

techniques. Circular dichroism (CD) spectroscopy is vital for monitoring the thermal stability of individual peptides and mixed peptide solutions, offering crucial insights into stability distinctions between homotrimers and heterotrimers [12, 13]. Nuclear magnetic resonance (NMR) plays an instrumental role in providing detailed information on individual residues and chains within trimer peptides, particularly in labeled complexes [14, 15]. Additionally, molecular dynamics (MD) simulation has significantly enhanced our understanding of how surface electrostatics and hydrogen bonding influence protein stability and fold specificity [16–18]. These sophisticated techniques collectively contribute to a comprehensive understanding of the structural and dynamic aspects of heterotrimeric peptides.

11.2 Covalently Tethering Heterotrimeric Peptides

11.2.1 Lys-Lys Covalent Linkage

The heterotrimeric structure of Type IV collagen is crucial for its binding with the $\alpha_1\beta_1$ integrin [19–22]. To mimic the composition of the critical binding site of Type IV collagen, Fields and colleagues devised a unique Lys-Lys covalent linkage of three peptide chains [23]. The synthesis of the heterotrimer, denoted as $[\alpha_1(\text{IV})]_2\alpha_2(\text{IV})$, entails an intricate multi-step process (Fig. 11.2). Initially, the α_2(IV) chain sequence was synthesized using Fmoc chemistry, where A represented Fmoc, B represented Dde, and the final Gly residue was introduced as Aloc-Gly. Subsequently, the Dde groups were removed, and Ahx was added, followed by incorporating the α_1(IV) sequence using Fmoc chemistry. In the third step, the Fmoc groups and Aloc group were eliminated, and (Gly-Pro-Hyp)$_6$ was added to each chain. Ultimately, the raw heterotrimer peptides were obtained after cleavage and underwent purification using RP-HPLC. This synthetic process necessitates meticulous control over chemical reactions to ensure the accurate production of the desired peptide sequences in the correct order.

11.2.2 Cys-Cys Disulfide Linkage

Covalent synthesis using disulfide bridging has emerged as a prominent strategy for constructing heterotrimeric collagen mimetic peptides. Typically, a single cysteine residue is incorporated into both the leading and lagging chains, while the middle chain requires the presence of two cysteine residues. The cysteines in the middle chain are protected by different groups, such as tert-butylthio (StBu) and acetamidomethyl (Acm), allowing for selective deprotection (Fig. 11.3A) [24]. This cysteine knot approach has been utilized to create heterotrimeric collagen-like peptides that accurately replicate collagenase cleavage sites and integrin binding sites.

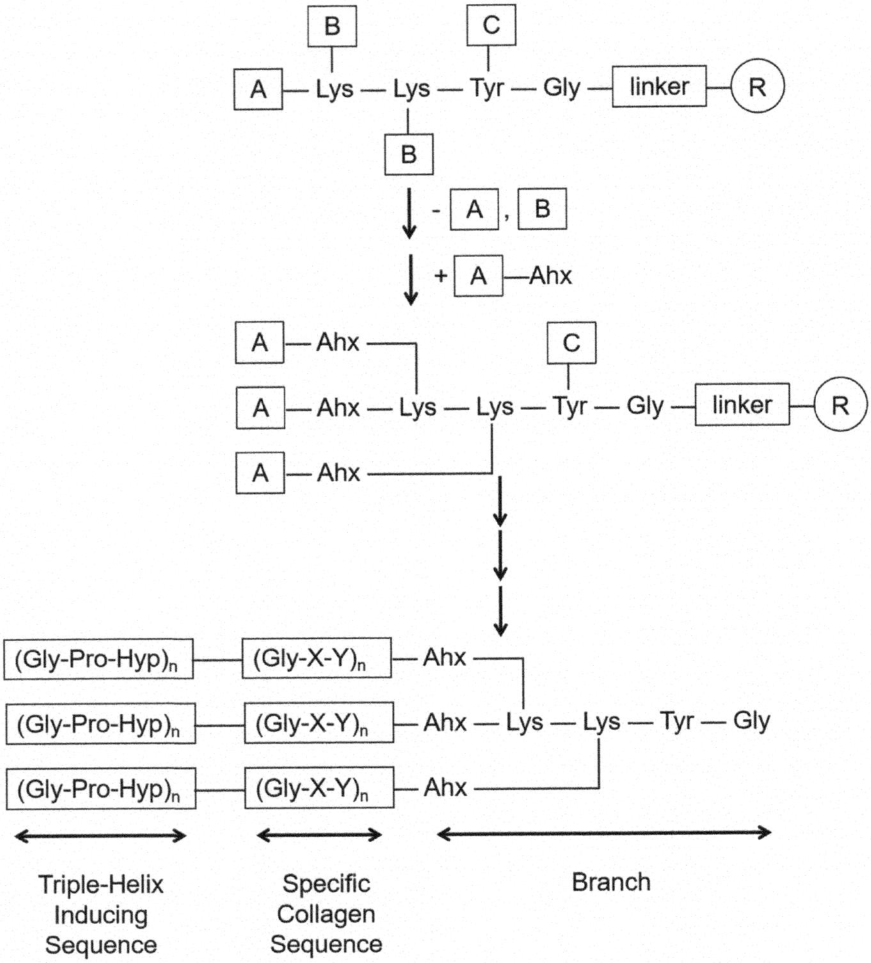

Fig. 11.2 A general scheme for the synthesis of heterotrimers with covalent linkage between Lys residues [23]. Reprinted from ref. [23], Copyright © 1996, with permission from ESCOM Science Publisher B.V

Heterotrimeric Type I collagen peptides, synthesized using the cysteine knot strategy, featured MMP-1 and MMP-8 cleavage sites in conjunction with an N-terminal nucleation domain (GPO)n (n = 3, 5) [5]. These meticulously designed peptides were then assembled into heterotrimers adopting the $\alpha_1\alpha_2\alpha_1'$ register (Fig. 11.3B). All the engineered heterotrimers exhibited stable triple-helical conformation, with their thermal stability varying between 9 and 41 °C, contingent on the length of the nucleation sequence. Notably, enzymatic digestion assays demonstrated efficient cleavage of the partially folded CMP by MMP-8, with the completely unfolded species exhibiting a slower cleavage process.

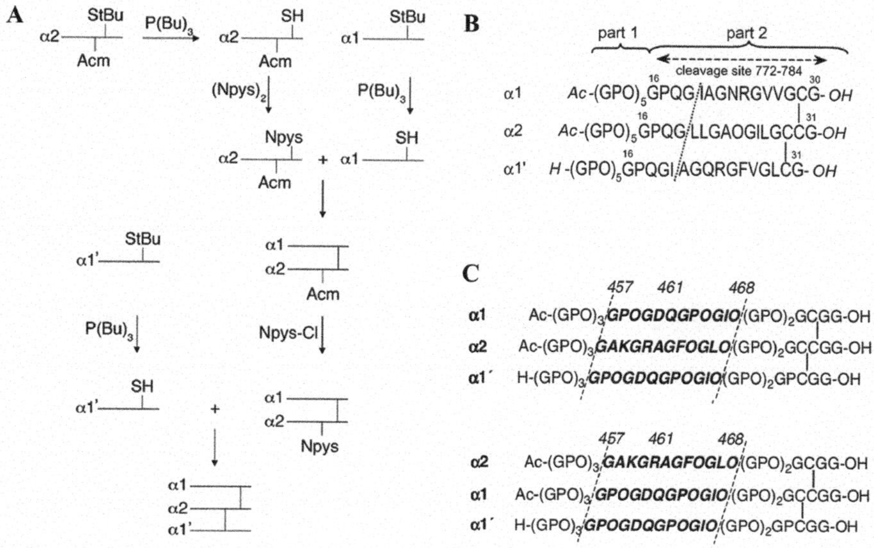

Fig. 11.3 Cys-Cys disulfide linkage for heterotrimeric CMP synthesis (**A**) [24] (Reprinted from ref. [24], Copyright © 2004, with permission from Wiley Periodicals, Inc.), collagenous heterotrimer that mimics the collagenase cleavage site of collagen Type I (**B**) [5] (Reprinted from ref. [5], Copyright © 2002, with permission from Elsevier Science Ltd.), and heterotrimeric collagen peptides containing the $\alpha 1\beta 1$ integrin recognition site (**C**) [25] (Reprinted from ref. [25], Copyright © 2002, with permission from European Peptide Society and John Wiley & Sons, Ltd.)

Heterotrimeric Type IV collagen peptides, incorporating integrin binding motifs flanked by $(GPO)_3$ at the N-terminus and $(GPO)_2$ at the C-terminus, were synthesized using the cysteine knot approach to evaluate their binding affinity to the $\alpha_2\beta_1$ integrin (Fig. 11.3C) [25]. The designed heterotrimeric peptides, adopting the $\alpha_1\alpha_2\alpha_1$' and $\alpha_2\alpha_1\alpha_1$' registers, formed stable triple helical structures, with respective T_m values of 42 and 30 °C, highlighting the crucial role of chain register in collagen stability. Notably, the heterotrimeric peptide with the $\alpha_2\alpha_1\alpha_1$' register exhibited slightly higher binding affinity to the $\alpha_2\beta_1$ integrin compared to the $\alpha_1\alpha_2\alpha_1$' register, providing evidence that the $\alpha_2\alpha_1\alpha_1$' register may represent the correct staggered alignment of the α subunits in Type IV collagen.

11.3 Charge Paired Heterotrimeric Peptides

11.3.1 Charge Paired AAB Heterotrimeric Peptides

Hartgerink and co-workers have achieved a significant milestone in synthesizing an AAB-type collagen heterotrimer through the effective utilization of electrostatic interactions between oppositely charged amino acids (Fig. 11.4). By designing two

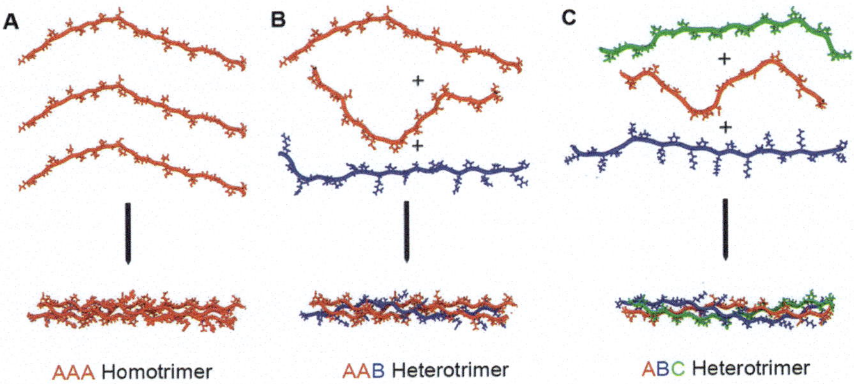

Fig. 11.4 Schematic diagram for the formation of charge paired heterotrimeric peptides ABC and AAB [11]. Reprinted from ref. [11], Copyright © 2010, with permission from Royal Society of Chemistry

peptides, $(EOGPOG)_5$ containing negatively charged Glu (denoted as A, $T_m = 46$ °C) and $(PRG)_{10}$ containing positively charged Arg, a stable heterotrimer with an impressive T_m value of 56 °C was formed upon mixing the two peptides in a 2:1 ratio [26]. Biophysical techniques, such as CD, differential scanning calorimetry (DSC), and NMR, confirmed that the AAB heterotrimer was the dominant species due to the favorable interaction between the Glu-Arg charged pairs. This study marks the first successful instance of a synthetic peptide system selectively producing a collagen heterotrimer with significantly enhanced stability compared to any of its homotrimeric components, representing a notable breakthrough in heterotrimer synthesis.

Hartgerink's group has further established a positive and negative design strategy to construct AAB collagen heterotrimers, effectively stabilizing the heterotrimer through favorable electrostatic interactions while destabilizing homotrimer formation [27]. The peptides $(PKGPOG)_5$ and $(EOG)_{10}$ were found to be incapable of forming homotrimers alone; however, when combined in a 2:1 ratio, they were able to form a register-specific heterotrimer peptide $(PKGPOG)_5{\cdot}(EOG)_{10}{\cdot}(PKGPOG)_5$, which exhibited a T_m value of 40 °C. The positive and negative design offers a robust approach for creating AAB heterotrimers, efficiently avoiding the formation of unwanted composite homotrimers.

11.3.2 Charge Paired ABC Heterotrimeric Peptides

Hartgerink's group developed a tripeptide system comprising a positively charged peptide A, a negatively charged peptide B and a neutral peptide C to fabricate ABC-type heterotrimer (Fig. 11.5A) [28]. Three peptides $(PKG)_{10}$, $(DOG)_{10}$ and $(POG)_{10}$ have been constructed, and their 1:1:1 mixture readily led to the formation of a stable

ABC heterotrimer with a T_m of 65 °C. Another peptide set consisting of $(PRG)_{10}$, $(EOG)_{10}$ and $(POG)_{10}$ has been revealed to form an ABC heterotrimer with a T_m of 54 °C. The main driving forces for these ABC heterotrimers are proposed to be salt-bridge and hydrogen bonding interactions between selectively constructed charged paired amino acids.

The positive and negative design strategy has been employed to create single-composition ABC collagen heterotrimers. Utilizing three peptides, $YG(PKGPKGPKGDKGPKG)_2$, $YG(DOG)_{10}$, and $YG(POGPOGPOGPOGPOG)_2$, in a 1:1:1 ratio, a highly stable ABC heterotrimer with a T_m value of 60 °C was formed. Additionally, the peptide $YG(EOG)_{10}$, denoted as B1, was explored and exhibited the formation of an AB1C heterotrimer with slightly reduced thermal stability (T_m = 52 °C) but enhanced specificity (Fig. 11.5B) [28]. This positive and negative design strategy offers an efficient means to avoid the formation of unwanted species while achieving the desired ABC heterotrimeric composition.

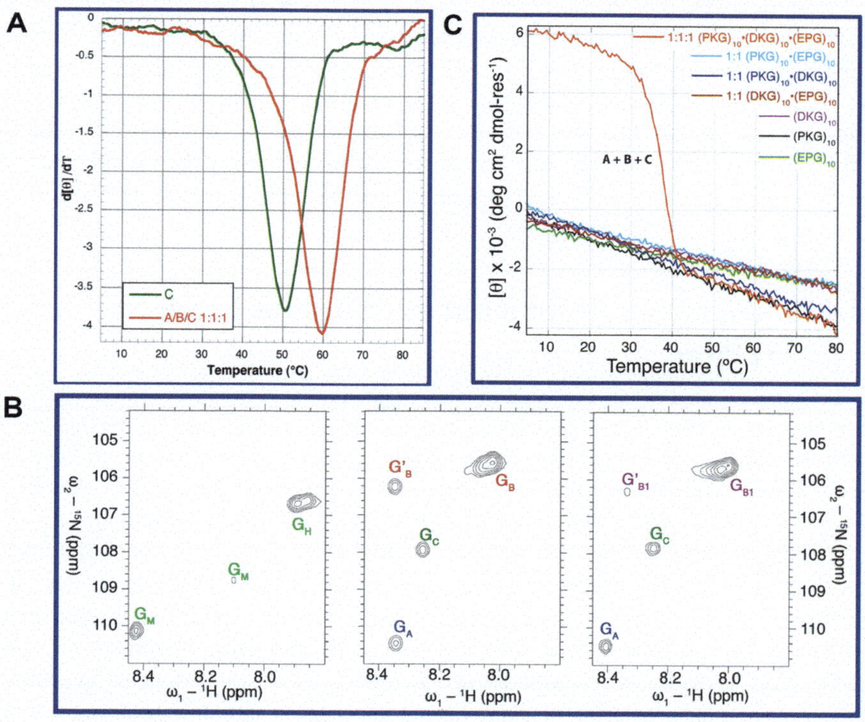

Fig. 11.5 Single-composition ABC collagen heterotrimers designed by the positive and negative design strategy (**A, B**) [28], (Reprinted from ref. [28], Copyright © 2012, with permission from American Chemical Society) and hydroxyproline-free single composition ABC collagen heterotrimer (**C**) [29]. (Reprinted from ref. [29], Copyright © 2013, with permission from American Chemical Society)

Moreover, three peptides, $Y(PKG)_{10}$, $Y(EPG)_{10}$ and $Y(DKG)_{10}$, were specifically designed to create a hydroxyproline-free ABC heterotrimer, inspired by bacterial collagen lacking hydroxyproline in its sequence [29]. Comprehensive CD and NMR analyses have confirmed the formation of a single-composition heterotrimer ABC with a T_m value of 37 °C, demonstrating that the absence of hydroxyproline does not compromise the thermal stability of the system (Fig. 11.5C). The remarkable thermal stability of the heterotrimer can be attributed to the abundant formation of salt-bridges between the positively charged Lys residues and the negatively charged Glu and Asp residues. This study introduces an innovative design concept for a collagen mimetic system that lacks hydroxyproline, with potential applications in expression systems.

Wennemers and collaborators recently harnessed the salt-bridge interaction between aminoproline (Amp) and aspartic acid (Asp) to meticulously guide the formation of ABC heterotrimers with well-defined composition and register (Fig. 11.6) [30]. CMP-1, featuring the sequence $(POG)_3$-Amp-Asp-Gly-$(POG)_3$, and its derivatives, incorporating the sequence $(POG)_3$-Xaa-Yaa-Gly-$(POG)_3$, were designed to probe the salt bridge interactions involving Amp-Asp (Fig. 11.6A). Remarkably, the formation of distinct ABC heterotrimers with high thermal stability (T_m value of 40 °C) was achieved with as few as three salt-bridges. Moreover, the exclusive assembly of the ABC heterotrimer occurred when Amp and Asp residues were paired in coplanar Xaa and Yaa positions, underscoring the significance of specific amino acid interactions in mediating heterotrimer formation (Fig. 11.6B).

11.4 Computational Optimization of Heterotrimeric Peptides

Computational simulation techniques have been investigated to refine the design of collagen-like heterotrimers by precisely manipulating their composition and register through optimization of pairwise interactions. Nanda and colleagues utilized a discrete computational model to design ABC-type heterotrimers, employing a positive and negative design approach to attain both stability and specificity in oligomerization (Fig. 11.7) [31]. Three 30-residue peptides (A, B, and C) were systematically optimized to favor Arg-Glu charge-pair interactions within an ABC heterotrimer, while simultaneously mitigating the formation of 26 competing oligomers. CD and NMR characterization of the peptides revealed that, despite the absence of a distinct A:B:C-type heterotrimer, the negative design approach resulted in the formation of A:B and B:C heterotrimers.

The substitution of Arg-Glu with Lys-Asp significantly improved the computational design of ABC-type collagen heterotrimer with specific compositions (Fig. 11.8) [32]. The optimization process aimed to maximize the energy gap between the desired state and competing misfolded states while maintaining a minimum level of target stability using automated computational techniques. CD analysis indicated

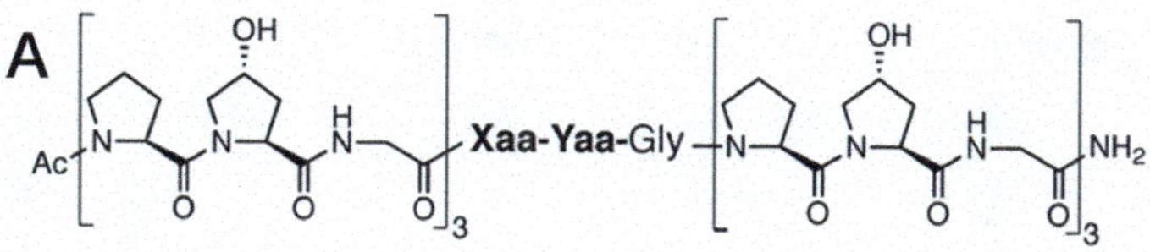

CMP	Xaa	Yaa	T_m (°C)
1	(4*S*)Amp	Asp	30
2	(4*S*)Amp	Pro	27
3	Pro	Asp	24
4	(4*S*)Amp	Glu	22

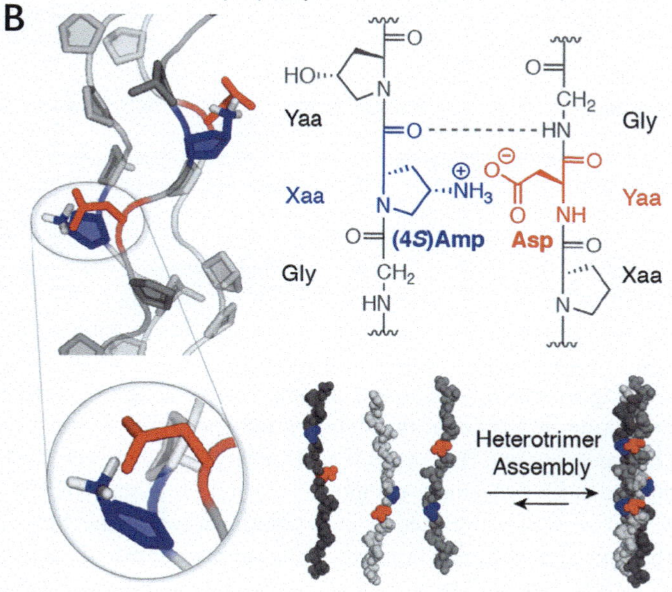

Fig. 11.6 A lateral salt-bridge for the specific assembly of an ABC-type collagen heterotrimer: peptide sequence (**A**) and salt-bridge analysis (**B**) [30]. Reprinted from ref. [30], Copyright © 2020, with permission from American Chemical Society

the essential requirement of all three peptides for the formation of a stable triple helix. Although the designed A:B:C heterotrimer exhibited enhanced thermodynamic stability compared to competing states, the computational approach did not account for the potential existence of multiple registers in the ABC heterotrimer.

Hartgerink and colleagues introduced an innovative scoring function to predict the stability of register-specific ABC heterotrimers, emphasizing the importance of axial Lys-Asp salt bridges in collagen stability (Fig. 11.9) [33]. This function assigned a value of 0 to a triple helical state resembling the $(POG)_{10}$ homotrimer, with negative values indicating destabilizing mutations and positive values signifying double mutations that meet Lys-Asp salt bridge requirements. Employing a genetic search

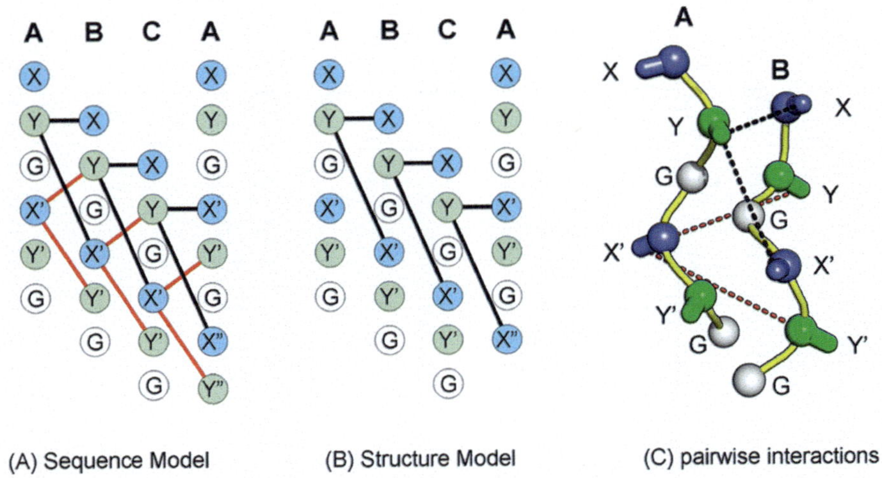

Fig. 11.7 De Novo designed collagen heterotrimers using positive and negative design [31]. Reprinted from ref. [31], Copyright © 2010, with permission from American Chemical Society

algorithm guided by this scoring function yielded optimized 30-residue peptide sequences, resulting in an ABC sequence with an outstanding score of −12. The designed ABC heterotrimer displayed exceptional stability, exhibiting a T_m of 58 °C, and precise control over its composition and register.

Xu and colleagues recently achieved an impressive design of synthetic ABC heterotrimers, taking into account the stability and specificity of axial and lateral salt bridges between Lys and Asp, as well as the energy gap between the target and 26 competing states [34]. A single-composition, single-register ABC heterotrimer with a T_m of 35 °C was accomplished by incorporating a complementary network of surface salt bridges and hydrogen bonding. The stability contributions were ranked as follows: axial KD > lateral DK > axial DK –lateral KD, with axial KD providing the highest contribution. However, complex salt-bridges involving multiple charged residues may exhibit anti-cooperativity due to geometric constraints imposed by the triple helix. Subsequently, the structure-based constraints in computational modeling have been innovatively devised to markedly refine the accuracy of chain register, thereby amplifying the stability and specificity of heterotrimers in complex systems.

11.5 Triblock Heterotrimeric Peptides

Conventional approaches to heterotrimer design often yielded less stable species compared to their homotrimer counterparts and imposed strict sequence requirements, making the incorporation of desired amino acids nearly impractical. To overcome these limitations, a novel strategy involving triblock peptides with three distinct domains has been developed, enabling the creation of AAB-type and ABC-type

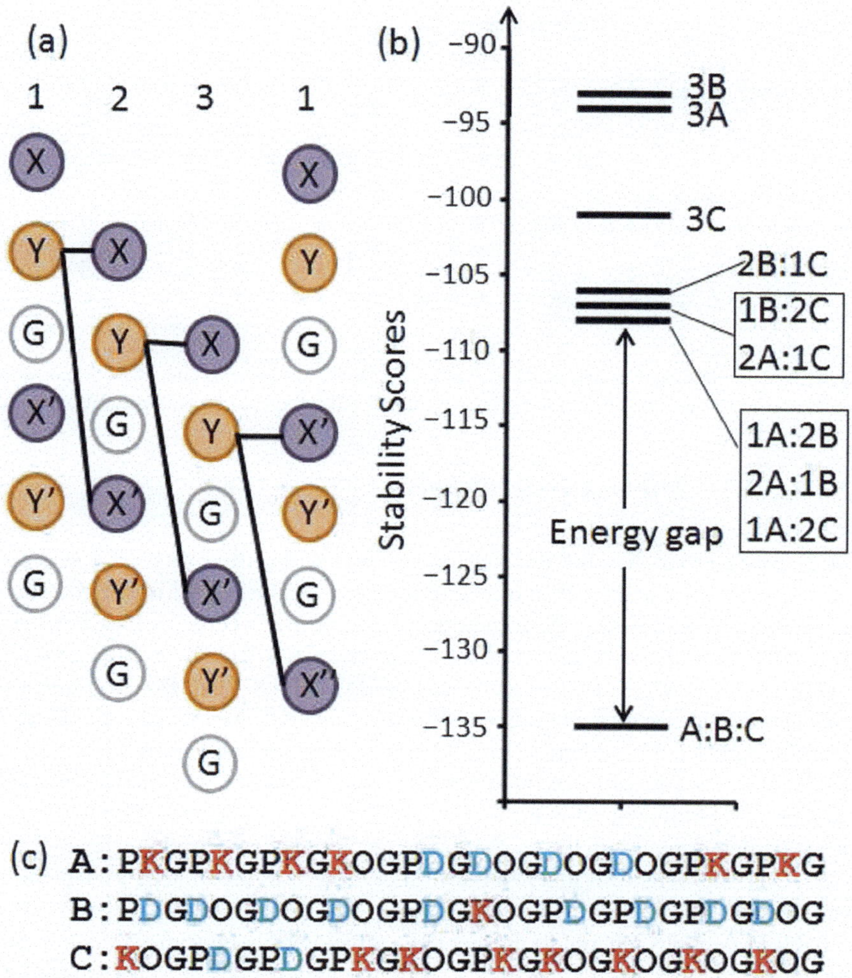

Fig. 11.8 Computational design of collagen A:B:C-type heterotrimers with controlled composition and multiple registers [32]. Reprinted from ref. [32], Copyright © 2011, with permission from American Chemical Society

heterotrimers that demonstrate enhanced stability and provide greater flexibility in terms of amino acid composition and arrangement. These triblock peptides not only open up new avenues for designing highly stable and customizable heterotrimeric structures but also enables the incorporation of functional motifs, significantly expanding potential applications.

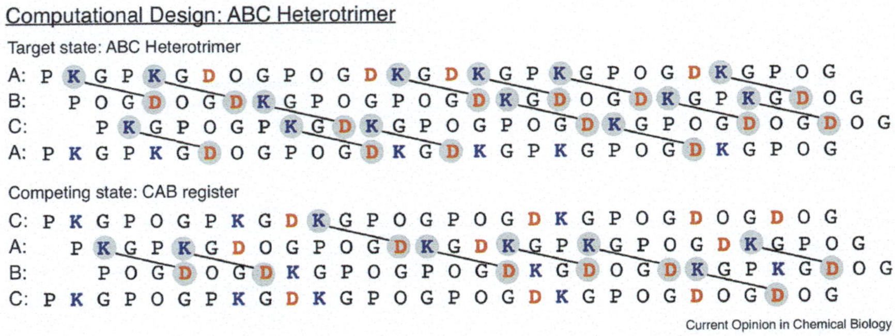

Fig. 11.9 Computational design of the collagen ABC heterotrimer with controlled composition and register [33]. Reprinted from ref. [33], Copyright © 2013, with permission from Elsevier Ltd.

11.5.1 AAB Triblock Heterotrimers

Our research group has for the first time reported the discovery of a series of triblock peptides capable of self-assembling into exclusive AAB heterotrimers, while completely avoiding the formation of homotrimers (Fig. 11.10). Two pairs of peptides: K-α1-D (with the amino acid sequence $K_5(GPO)_3GFOGER(GPO)_3D_5$) and D-α1-K (with the amino acid sequence $D_5(GPO)_3GFOGER(GPO)_3K_5$) as well as K-α2-D (with the amino acid sequence $K_5(GPO)_3GPOGES(GPO)_3D_5$) and D-α2-K (with the amino acid sequence $D_5(GPO)_3GPOGES(GPO)_3K_5$) have been synthesized to drive the formation of AAB heterotrimers, respectively. Notably, GFOGER is the integrin binding motif in the α_1 chain of Type I collagen, while GPOGES is its corresponding sequence in the α_2 chain. The recombination of peptides K-α1-D and D-α2-K, as well as K-α2-D and D-α1-K further showcases the strategy's adaptability in mimicking the natural composition of Type I collagen.

Triblock peptides inherently prevent the formation of homotrimers due to repulsion between like-charged amino acids. Combining two oppositely charged triblock peptides results in the assembly of stable heterotrimers with a T_m value of approximately 30 °C, facilitated by attractive forces between complementarily charged amino acids. Importantly, the central collagen-like block of the triblock peptides allows for the convenient incorporation of various desired amino acids without hindering heterotrimer formation. NMR analysis demonstrates that the composition of the heterotrimer can be easily controlled by adjusting the relative concentration of peptides A and B. Molecular dynamics simulations further illustrate that the extensive salt-bridge network formed by the oppositely charged terminal amino acids significantly contributes to the stability of the heterotrimers.

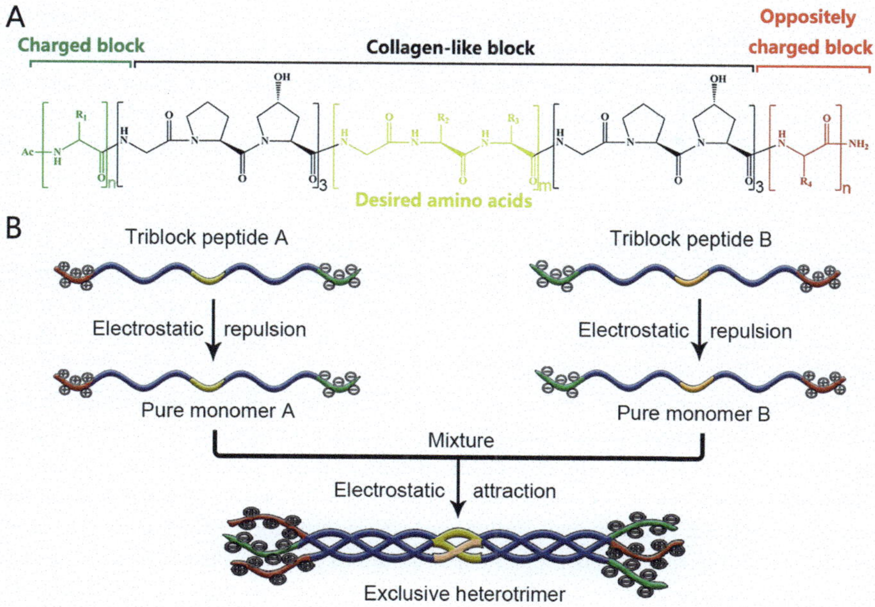

Fig. 11.10 Schematic illustration of triblock peptides to form heterotrimer. Amino acid sequences of triblock peptides (**A**). The self-assembly of triblock peptides into exclusive heterotrimer (**B**)

11.5.2 ABC Triblock Heterotrimeric Peptides

Our group has achieved further progress in developing a tripeptide system, comprising triblock peptides A, B, and C, to facilitate the formation of stable ABC-type heterotrimers. Peptide A ($G_3(GPO)_3GLOGFK(GPO)_3K_3$) contains a charged C-terminal block, peptide B ($D_3(GPO)_3GLSGVO(GPO)_3G_3$) possesses an oppositely charged N-terminal block, and peptide C ($K_3(GPO)_3GFOGER(GPO)_3D_3$) features oppositely charged N/C-terminal blocks. Combining these triblock peptides in a 1:1:1 ratio readily forms a stable ABC heterotrimer with a T_m of 24 °C. The repulsion between same-charged amino acids prevents the formation of homotrimers, while the attractive forces between the oppositely charged peptides drive the assembly of the heterotrimer.

11.6 Covalent Capture of Heterotrimeric Peptides

The covalent capture of oppositely charged amino acids has been established as a means to stabilize the triple-helical structure of heterotrimers [35, 36]. These amino acids not only drive the self-assembly of register-specific heterotrimers but also provide covalent linking sites. Peptides A (with the sequence GPPGPPGPKGDPGPKGDPGPKGDPGPPGGY) and B (with the sequence GPPGPKGPPGDKGPPGDKGPPGDPGPPGGY) were designed to form a stable AAB heterotrimer (named KGD) through K-D charge-pair interactions (Fig. 11.11A) [37]. To further enhance stability, peptide KGD was covalently captured by proximity-directed amide bond formation between K and D using EDC/HOBT activation, resulting in the formation of peptide KGDcc. Peptide KGDcc demonstrated a significant increase in T_m values compared to peptide KGD, with a difference of +19.5 °C (T_m of 45.5 °C for KGD and 67.0 °C for KGDcc). Additionally, peptide KGDcc displayed a much faster folding rate, reaching 100% folded within 25 min, whereas peptide KGD took 5.5 h, indicating that the presence of covalent bonds greatly enhanced the refolding capability of heterotrimeric peptides.

The covalent capture of collagen triple helices through K-E isopeptide bond formation has emerged as a more efficient approach compared to K-D covalent capture (Fig. 11.11B) [38]. The K-E covalent capture process exhibits faster kinetics, with nearly complete yield, while the K-D reaction does not reach full completion. Moreover, the K-E covalent capture process significantly enhances the thermal stability, refolding rate, and triple helical structure of the heterotrimers, whereas the K-D covalent capture process does not have the same impact on the triple helical structure. These differences can be attributed to the additional methylene group in the side chain of Glu, which confers greater flexibility to adopt the triple helical conformation.

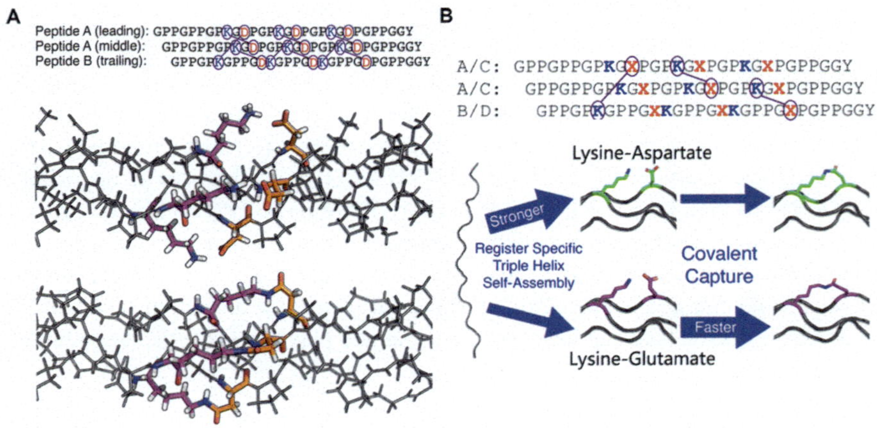

Fig. 11.11 Covalent capture of heterotrimeric collagen triple helices using lysine – aspartate (**A**) [37] (Reprinted from ref. [37], Copyright © 2019, with permission from American Chemical Society), and lysine – glutamate pairs (**B**) [38] (Reprinted from ref. [38], Copyright © 2020, with permission from American Chemical Society), X represents aspartate in the case of A/B peptides and glutamate in the case of C/D peptides

11.7 Concluding Remarks

The evolving field of collagen heterotrimeric peptide construction is fueled by the pursuit of understanding their functions and harnessing their potential applications. Diverse strategies, encompassing covalent and non-covalent approaches, have emerged through extensive research. Non-covalent self-assembly methods, particularly those utilizing charge-pair interactions, computational optimization, and triblock peptides, have gained prominence for their simplicity, scalability, and versatility. These innovations revolutionize the precise control over heterotrimer formation, enabling tailored composition and register. These advancements not only deepen our knowledge of collagen-like peptides but also open new avenues across fields, from regenerative medicine to biomaterials. Continued exploration and refinement of these strategies are poised to lead to further breakthroughs and exciting applications in the future.

Despite substantial progress in noncovalent self-assembled peptides, the challenge remains to achieve stable heterotrimeric systems with desired functional motifs. Advancing the field of collagen heterotrimers necessitates novel strategies to attain highly stable heterotrimeric systems with minimal interfering homotrimers. The development of versatile heterotrimeric peptides with specific functional motifs holds the potential to significantly expand their application scope. Additionally, exploring higher-order heterotrimer biomaterials presents exciting opportunities for creating complex structures with enhanced functionality. These endeavors promise to deepen our understanding of collagen's function and pathogenesis while unlocking new possibilities in regenerative medicine and biomaterial engineering.

References

1. Moremen KW, Tiemeyer M, Nairn AV (2012) Vertebrate protein glycosylation: diversity, synthesis and function. Nat Rev Mol Cell Biol 13:448–462
2. Ottl J, Gabriel D, Murphy G, Knäuper V, Tominaga Y, Nagase H et al (2000) Recognition and catabolism of synthetic heterotrimeric collagen peptides by matrix metalloproteinases. Chem Biol 7:119–132
3. Ricard-Blum S (2011) The collagen family. Cold Spring Harb Perspect Biol 3:a004978
4. Sorushanova A, Delgado LM, Wu Z, Shologu N, Kshirsagar A, Raghunath R et al (2019) The collagen suprafamily: from biosynthesis to advanced biomaterial development. Adv Mater 31:e1801651
5. Fiori S, Sacca B, Moroder L (2002) Structural properties of a collagenous heterotrimer that mimics the collagenase cleavage site of collagen type I. J Mol Biol 319:1235–1242
6. Jalan AA, Sammon D, Hartgerink JD, Brear P, Stott K, Hamaia SW et al (2020) Chain alignment of collagen I deciphered using computationally designed heterotrimers. Nat Chem Biol 16:423–+
7. Fields CG, Lovdahl CM, Miles AJ et al (1993) Solid-phase synthesis and stability of triple-helical peptides incorporating native collagen sequences. Biopolymers: Original Res Biomol 33(11):1695–1707

8. Ottl J, Battistuta R, Pieper M et al (1996) Design and synthesis of heterotrimeric collagen peptides with a built-in cystine-knot Models for collagen catabolism by matrix-metalloproteases. FEBS Lett 398(1):31–36
9. Ottl J, Moroder L (1999) Disulfide-bridged heterotrimeric collagen peptides containing the collagenase cleavage site of collagen type I. Synthesis and conformational properties. J Am Chem Soc 121(4):653–661
10. Ottl J, Moroder L (1990) A new strategy for regioselective interstrand disulfide bridging of multiple cysteine peptides. Tetrahedron Lett 40(8):1487–1490
11. Fallas JA, O'Leary LER, Hartgerink JD (2010) Synthetic collagen mimetics: self-assembly of homotrimers, heterotrimers and higher order structures. Chem Soc Rev 39:3510–3527
12. Fuller A, Wall A, Crowther MD et al (2017) Thermal stability of heterotrimeric pMHC proteins as determined by circular dichroism spectroscopy. Bio-Protoc 7(13):e2366–e2366
13. Koide T, Nishikawa Y, Takahara Y (2004) Synthesis of heterotrimeric collagen models containing Arg residues in Y-positions and analysis of their conformational stability. Bioorg Med Chem Lett 14(1):125–128
14. Madhan B, Xiao J, Thiagarajan G et al (2008) NMR monitoring of chain-specific stability in heterotrimeric collagen peptides. J Am Chem Soc 130(41):13520–13521
15. Xiao J, Sun X, Madhan B et al (2015) NMR studies demonstrate a unique AAB composition and chain register for a heterotrimeric type IV collagen model peptide containing a natural interruption site. J Biol Chem 290(40):24201–24209
16. Singam ERA, Balamurugan K, Gopalakrishnan R et al (2012) Molecular dynamic simulation studies on the effect of one residue chain staggering on the structure and stability of heterotrimeric collagen-like peptides with interruption. Biopolymers 97(11):847–863
17. Gautieri A, Vesentini S, Montevecchi FM et al (2008) Mechanical properties of physiological and pathological models of collagen peptides investigated via steered molecular dynamics simulations. J Biomech 41(14):3073–3077
18. Bodian DL, Radmer RJ, Holbert S et al (2011) Molecular dynamics simulations of the full triple helical region of collagen type I provide an atomic scale view of the protein's regional heterogeneity. Biocomputing 2011:193–204
19. Saccà B, Fiori S, Moroder L (2003) Studies of the local conformational properties of the cell-adhesion domain of collagen type IV in synthetic heterotrimeric peptides. Biochemistry 42(12):3429–3436
20. Glanville R W (1987) Type IV collagen. Structure and function of collagen types 43.
21. Wu Y, Ge G (2019) Complexity of type IV collagens: from network assembly to function. Biol Chem 400(5):565–574
22. Odermatt BF, Lang AB, Rüttner JR et al (1984) Monoclonal antibodies to human type IV collagen: useful reagents to demonstrate the heterotrimeric nature of the molecule. Proc Natl Acad Sci 81(23):7343–7347
23. Fields CG, Grab B, Lauer JL, Miles AJ, Yu YC, Fields GB (1996) Solid-phase synthesis of triple-helical collagen-model peptides. Lett Pept Sci 3:3–16
24. Renner C, Sacca B, Moroder L (2004) Synthetic heterotrimeric collagen peptides as mimics of cell adhesion sites of the basement membrane. Biopolymers 76:34–47
25. Sacca B, Moroder L (2002) Synthesis of heterotrimeric collagen peptides containing the alpha 1 beta 1 integrin recognition site of collagen type IV. J Pept Sci 8:192–204
26. Russell LE, Fallas JA, Hartgerink JD (2010) Selective assembly of a high stability AAB collagen heterotrimer. J Am Chem Soc 132:3242–+
27. O'Leary LE, Fallas JA, Hartgerink JD (2011) Positive and negative design leads to compositional control in AAB collagen heterotrimers. J Am Chem Soc 133:5432–5443
28. Fallas JA, Lee MA, Jalan AA, Hartgerink JD (2012) Rational design of single-composition ABC collagen heterotrimers. J Am Chem Soc 134:1430–1433
29. Jalan AA, Demeler B, Hartgerink JD (2013) Hydroxyproline-free single composition ABC collagen heterotrimer. J Am Chem Soc 135:6014–6017
30. Hentzen NB, Islami V, Kohler M, Zenobi R, Wennemers H (2020) A lateral salt bridge for the specific assembly of an ABC-type collagen heterotrimer. J Am Chem Soc 142:2208–2212

31. Xu F, Zhang L, Koder RL, Nanda V (2010) De novo self-assembling collagen heterotrimers using explicit positive and negative design. Biochemistry 49:2307–2316
32. Xu F, Zahid S, Silva T, Nanda V (2011) Computational design of a collagen A:B:C-type heterotrimer. J Am Chem Soc 133:15260–15263
33. Jalan AA, Hartgerink JD (2013) Pairwise interactions in collagen and the design of heterotrimeric helices. Curr Opin Chem Biol 17:960–967
34. Zheng H, Lu C, Lan J, Fan S, Nanda V, Xu F (2018) How electrostatic networks modulate specificity and stability of collagen. Proc Natl Acad Sci U S A 115:6207–6212
35. Kubyshkin V (2019) Stabilization of the triple helix in collagen mimeticking peptides. Org Biomol Chem 17(35):8031–8047
36. Xu Y, Kirchner M (2021) Collagen mimetic peptides. Bioengineering 8(1):5
37. Li IC, Hulgan SAH, Walker DR, Farndale RW, Hartgerink JD, Jalan AA (2019) Covalent capture of a heterotrimeric collagen helix. Org Lett 21:5480–5484
38. Hulgan SAH, Jalan AA, Li IC, Walker DR, Miller MD, Kosgei AJ et al (2020) Covalent capture of collagen triple helices using lysine-aspartate and lysine-glutamate pairs. Biomacromol 21:3772–3781

Chapter 12
Self-assembly of Collagen Mimetic Peptides

Abstract Collagen mimetic peptides (CMPs) present promising synthetic alternatives to animal-derived collagen, boasting advantages like reduced risk of viral infection, low immunogenicity, and convenient quality control. However, mimicking collagen's intricate self-assembly process in short CMPs averaging only 10 nm presents significant challenges. This chapter delves into strategies for devising the self-assembly of short CMPs into homotrimers and heterotrimers to replicate collagen's structural characteristics. It thoroughly examines strategies for homotrimer peptide self-assembly, including diverse approaches such as amphiphilic interactions, π–π interactions, metal ion coordination, electrostatic interactions, triblock peptides, and covalent cross-linking, facilitating the creation of well-defined supramolecular architectures at the nano- and micro-scale, encompassing structures like fibers, microflorettes, meshes, microtubes, hollow spheres, nanodisks, and nanosheets. Additionally, innovative strategies involving covalent cross-linking, electrostatic interactions, and Ln^{3+}-ligand interactions are explored to promote the self-assembly of heterotrimeric peptides into supramolecular structures. The chapter concludes by highlighting applications of self-assembled collagen mimetic peptides in cargo delivery, cell culture, and hemostatic applications, emphasizing their potential in various biomedical fields.

Keywords Self-assembly · Self-assembled homotrimer peptides · Self-assembled heterotrimer peptides

12.1 Introduction to Self-assembled Collagen Mimetic Peptides

Collagen has become a focal point for diverse biomedical applications, due to its distinctive hierarchical structures and versatile biological functions [1–3]. The repetitive $(Gly\text{-}X\text{-}Y)_n$ triplets uniquely enable collagen to assemble into triple helices, which, in turn, self-assemble into D-periodic fibrils. These fibrils interconnect to form a molecular scaffold, imparting structural integrity and mechanical stability

© Springer Nature Singapore Pte Ltd. 2024

J. Xiao, *Collagen Mimetic Peptides and Their Biophysical Characterization*,
Springer Series in Biomaterials Science and Engineering 20,
https://doi.org/10.1007/978-981-10-1941-8_12

to connective tissues [4, 5]. Furthermore, collagen engages with a myriad of matrix molecules and cellular receptors, influencing various cellular processes such as adhesion, migration, and development [6–9]. Collagen-based materials have demonstrated broad applications in tissue engineering and regenerative medicine [10–15]. However, concerns regarding potential pathogen transmission associated with animal-derived collagen underscore the need to explore alternative collagen sources.

Collagen mimetic peptides (CMPs) have emerged as promising synthetic alternatives to animal-derived collagen, offering key benefits such as a diminished risk of viral infection, low immunogenicity, and convenient quality control. Despite their potential, mimicking the intricate self-assembly process of collagen in short CMPs with an average length of only 10 nm poses considerable challenges. Recent advancements have focused on devising strategies for the self-assembly of short CMPs into homotrimers and heterotrimers, mimicking the structural characteristics of collagen. While homotrimer self-assembly has been extensively investigated, the exploration of heterotrimeric peptide self-assembly remains in its nascent stages. Nonetheless, self-assembled CMPs have showcased considerable promise across diverse applications, including cargo transportation, cellular cultivation, and hemostasis. Ongoing research and innovation in this domain are poised to catalyze the development of advanced collagen-inspired materials with enhanced functionalities.

12.2 Self-assembled Homotrimer Peptides

Homotrimer peptides, comprised of three identical α chains, serve as versatile building blocks for the self-assembly of collagen-like structures, employing innovative strategies such as amphiphilic interactions, π–π stacking, metal–ligand coordination, electrostatic interactions, and covalent cross-linking. Homotrimer peptides have facilitated the creation of diverse and well-defined supramolecular architectures at the nano- and micro-scale, encompassing structures such as fibers, microflorettes, meshes, microtubes, hollow spheres, nanodisks, and nanosheets, presenting promising opportunities for the development of biomaterials with customized properties.

12.2.1 Amphiphilic Interactions

Amphiphilic peptides have demonstrated remarkable self-assembly capabilities, enabling the formation of a diverse range of supramolecular structures. Tong and colleagues introduced a collagen-mimetic peptide amphiphile (CPA) consisting of four segments: a hydrophobic tail C16, a β-sheet segment, a spacer, and a collagen-like peptide with the bioactive motif GFOGER [16]. The CPA adopted triple-helical conformation in aqueous solution and self-assembled into collagen-like nanofibers upon screening the positive charges of the lysine spacer (Fig. 12.1).

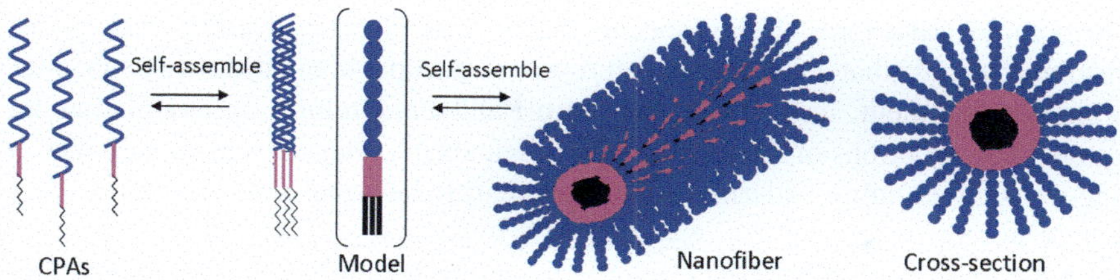

Fig. 12.1 Schematic illustration of the self-assembly of amphiphilic CMPs: peptides with C16 [16]. Reprinted from ref. [16], Copyright © 2011, with permission from American Chemical Society

Our research group recently reported the development of a novel class of amphiphile-like peptides that conjugated repetitive triple-helical $(GPO)_m$ sequences with C-terminal hydrophilic aspartic acids. The amphiphile-like CMPs with variable length of $(GPO)_m$ sequences consistently self-assembled into well-ordered nanospherical supramolecular structures [17]. Notably, the number of terminal aspartic acids was found to modulate the morphology of assembled amphiphile-like CMPs. A series of bolaamphiphile-like peptides consisting of two hydrophilic heads with charged aspartic acids and a central collagen-like triple helical domain including $(GPP)_m$ or $(GPO)_m$ has been further explored, indicating that the presence of terminal aspartic acids plays a crucial role in inducing the self-assembly of these collagen mimetic peptides into nanospheres [18].

12.2.2 π–π Interactions

Maryanoff and colleagues elucidated the pivotal role of π–π interactions, facilitated by aromatic amino acids such as phenylalanine (Phe), in directing the self-assembly of collagen mimetic peptides (CMPs). The incorporation of electron-deficient pentafluorophenylalanine (F5-Phe) and Phe at the N- and C-terminal ends of the CMPs has been observed to result in the formation of micrometer-long composite fibrils with a diameter of 0.26 μm. Subsequent modifications of aromatic residues unveiled the influence of π–π interactions on the morphology of self-assembled structures, ranging from fibrils with branches to spherical aggregates and sheets with irregularly shaped holes [19, 20]. These findings underscore the critical role of π–π interactions in guiding the appropriate self-assembly of CMPs.

12.2.3 Metal Ion Coordination

Metal ions have emerged as a promising external stimulus for regulating peptide self-assembly due to their unique ability to selectively and reversibly bind to peptide ligand groups, enabling programmable and hierarchical self-assembly of peptides into complex structures. Using specific metal ions offers a flexible strategy for crafting materials that are dynamic and responsive, capable of adjusting to changes in their environment, such as variations in pH and temperature. By modulating the types and distributions of the metal-binding ligands in collagen mimetic peptides, a range of self-assembly patterns, including linear, radial, and cross formations, has been achieved.

Chmielewski and colleagues developed a peptide named NCoH, which integrates a metal-binding ligand, nitrilotriacetic acid (NTA), and two histidine residues at both ends of a $(POG)_9$ peptide. This structure, upon interaction with divalent metal ions, forms distinct microflorettes, averaging 10–20 μm in diameter (Fig. 12.2A) [21]. They also crafted IdaCol and HisCol, two triple helical peptides, with IdaCol featuring iminodiacetic acid and HisCol having two histidine residues at both terminals of $(POG)_9$, leading to metal-induced linear co-assembly into petal-like structures with 9–12 nm periodic gaps when exposed to divalent metal ions (Fig. 12.2B) [22]. These results demonstrate that the incorporation of metal-binding ligands, such as histidines or iminodiacetic acid, at both terminals of collagen mimetic peptides facilitates their linear self-assembly in the presence of divalent metal ions.

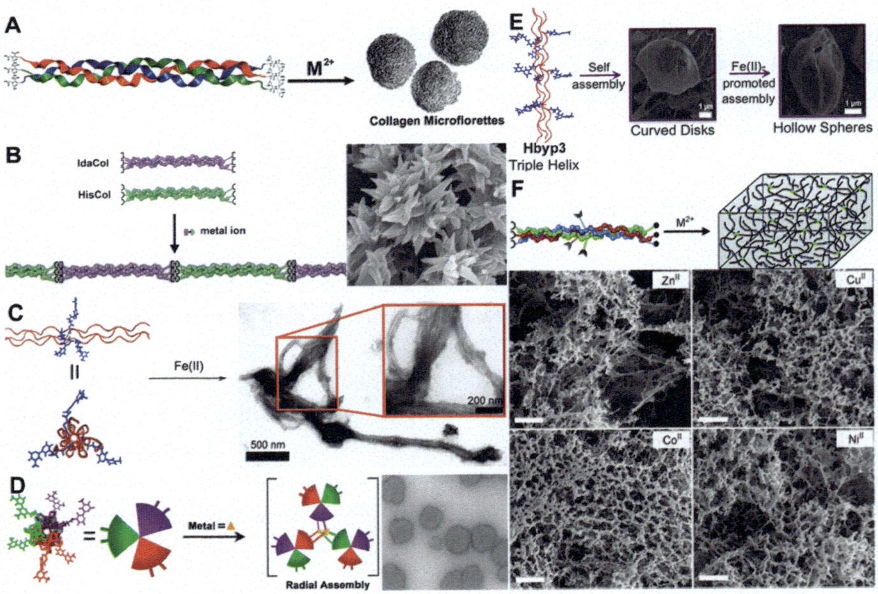

Fig. 12.2 Schematic illustration of self-assembled CMPs mediated by divalent metal ions and their corresponding TEM and SEM images: linear assembled peptides (**A**, **B**) [21, 22] (Reprinted from refs. [21, 22], Copyright © 2009 and 2011, with permission from American Chemical Society), radial assembled peptides (**C**, **D**, **E**) [23–25] (Reprinted from refs. [23, 25], Copyright © 2008, 2010 and 2013, with permission from American Chemical Society), and cross-assembled peptides (**F**) [26] (Reprinted from ref. [26], Copyright © 2009, with permission from WILEY–VCH Verlag GmbH & Co. KGaA, Weinheim)

Chmielewski's team expanded their work by creating radial peptides, infusing various metal ion-binding ligands into the $(POG)_9$ peptide sequence. They replaced a central hydroxyproline (Hyp) with a bipyridine-modified lysine, leading to the self-assembly of the peptide into fibers with Fe^{2+} ions (Fig. 12.2C) [23]. Altering two additional Hyp residues with bipyridine-modified lysines at specific positions resulted in the formation of disc-like structures with diameters ranging from 50 to 500 nm upon exposure to Fe^{2+} ions (Fig. 12.2D) [24]. Interestingly, a variant with three bipyridine-modified lysines spontaneously formed curved disks, and in the presence of Fe^{2+} ions, transformed into hollow spheres (Fig. 12.2E) [25].

Moreover, they have innovated a cross-assembled peptide named HBN by integrating metal-binding ligands NTA and bpy, along with two histidines (His2), at the N-terminal, central position, and C-terminal of the $(POG)_9$ peptide, respectively. When exposed to different metal ions, the HBN peptide showed a unique ability to form complex, cross-linked 3D fibrous scaffolds, with their structure varying based on the specific metal ion used (Fig. 12.2F) [26]. For instance, scaffolds formed with Zn^{2+} were less dense compared to those with Co^{2+}, Cu^{2+}, and Ni^{2+}. The interplay between transition metal ions and ligands presents a dynamic strategy for creating a diverse array of supramolecular structures.

Beyond heavy metal ions, our research group has explored lanthanide ions for directing the self-assembly of collagen mimetic peptides (CMPs). Lanthanides offer benefits like low toxicity, high biocompatibility, and pronounced luminescence. The innovative peptide $DD(PPG)_{12}DD$, upon interaction with lanthanide ions, uniquely assembles into helical nanoropes with noticeable periodic banding (Fig. 12.3A) [27]. Additionally, three collagen mimetic peptides (CMPs) $DD(GPO)_7DD$, $DW(GPO)_7WD$, and $HW(GPO)_7WD$ have exhibited the capability to self-assemble into biocompatible nanofibers when activated by various lanthanide ions, enabling the functionalization of these materials with easily adjustable photoluminescence (Fig. 12.3B) [28].

A range of lanthanide ion-mediated collagen peptides, named $DD(GPO)_5$-Z-$PO(GPO)_4DD$ (Z = Ala, Arg, Asp, Cys, Glu, Ser, and Val), were developed to explore the impact of Glycine (Gly) mutations on their structure and assembly. These peptides were identified by the residue replacing Gly, with the unmutated version termed peptide G. Below their melting temperature, peptides G, A, S, and C created well-defined triple helices and organized fibrils, whereas near the critical temperature, G, A, S, C, V, R, E, and D transformed into a mix of triple helices and monomers, forming disordered aggregates, and above this temperature, all remained as unfolded monomers, incapable of assembly [29]. This detailed study sheds light on how Gly mutations affect collagen peptide self-assembly.

12.2.4 Electrostatic Interactions

Electrostatic interactions between oppositely charged amino acids have been leveraged to drive the self-assembly of CMPs. Chaikof and Conticello developed

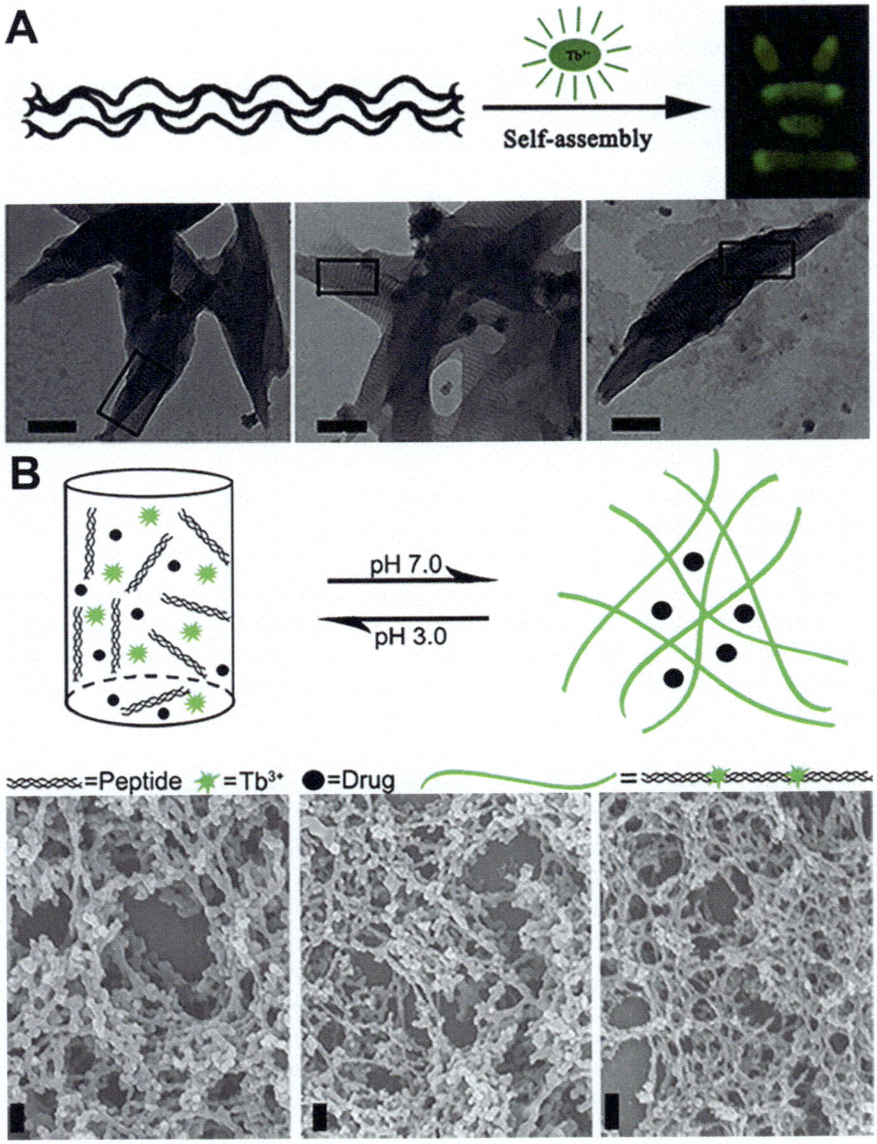

Fig. 12.3 Schematic illustration of lanthanide ion-mediated CMPs: peptide nanoropes (**A**) [27] (Reprinted from ref. [27], Copyright © 2016, with permission from WILEY–VCH Verlag GmbH & Co. KGaA, Weinheim), and peptide framework (**B**) [28] (Reprinted from ref. [28], Copyright © 2019, with permission from American Chemical Society)

the CPII peptide, featuring the sequence (Pro-Arg-Gly)$_4$(Pro-Hyp-Gly)$_4$(Glu-Hyp-Gly)$_4$, which demonstrated the ability to self-assemble into fibrils showing distinct D-banding (Fig. 12.4A, B) [30]. This process entailed the peptides first coming together to form blunt-ended triple helices, which then aligned in a staggered configuration. Expanding on this model, Conticello's group developed 2D nanosheets by orderly assembling two peptides, NSI and NSII, substituting Arginine with 4-aminoproline (Fig. 12.4C) [31].

Hartgerink and colleagues advanced the design of peptide CPII by developing peptide KOD, featuring the sequence (Pro-Lys-Gly)$_4$(Pro-Hyp-Gly)$_4$(Asp-Hyp-Gly)$_4$. This peptide displayed a distinctive self-assembly process, starting from forming triple helices, evolving into nanofibers, and culminating in the creation of a hydrogel (Fig. 12.4D) [32]. This pathway marks a departure from the conventional "blunt-ended" approach, embodying a "sticky-ended" assembly technique.

In pursuit of a deeper understanding of the self-assembly mechanism, four peptides F1 [(PKG)$_4$(DOG)$_4$(POG)$_4$], F2 [(POG)$_4$(PKG)$_4$(DOG)$_4$], A1 [(DOG)$_4$(PKG)$_4$(POG)$_4$] and A2 [(POG)$_4$(DOG)$_4$(PKG)$_4$] were designed by rearranging the domain positions originally found in peptide KOD [33]. Surprisingly, despite the "blunt-ended" assembly expectations suggesting that none of these peptides should form fibers due to the presence of a single terminal with a charged domain, transmission electron microscopy (TEM) images revealed fiber formation for peptides F1 and F2, while peptides A1 and A2 formed amorphous aggregates. These results significantly support the "sticky-ended" self-assembly mechanism in these peptides, challenging conventional assumptions.

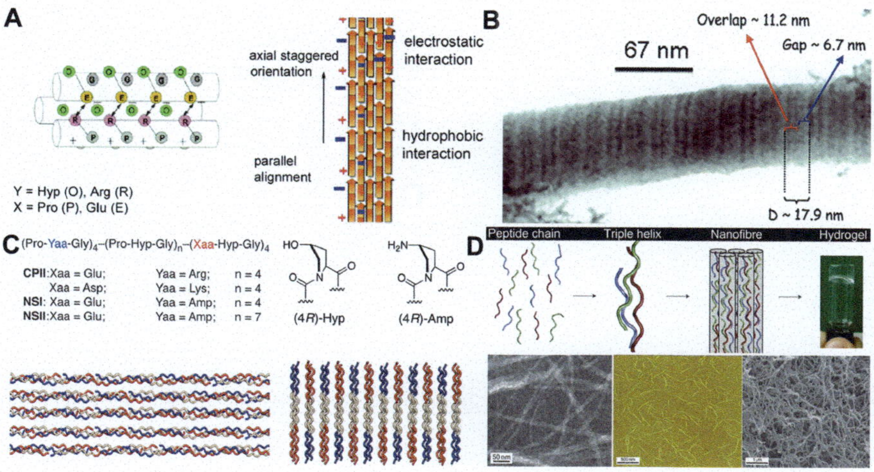

Fig. 12.4 Schematic illustration of electrostatic interactions-triggered CMPs: D-periodic collagen-mimetic microfibers (**A, B**) [30] (Reprinted from ref. [30], Copyright © 2007, with permission from American Chemical Society), multilayer collagen nanosheets (**C**) [31] (Reprinted from ref. [31], Copyright © 2014, with permission from American Chemical Society), and self-assembled collagen hydrogels (**D**) [32] (Reprinted from ref. [32], Copyright © 2011, with permission from Springer Nature Limited)

12.2.5 Triblock Collagen Mimetic Peptides

Triblock copolymers, known for their convenient modulation of hydrophobic/hydrophilic blocks, have gained substantial recognition as versatile materials with applications in drug delivery, thermoplastic elastomers, and new energy materials. Our research group has achieved a groundbreaking milestone by creating a series of triblock peptides composed solely of natural amino acids, which spontaneously assemble into collagen-like fibers displaying a distinct D-banding pattern (Fig. 12.5). These peptides comprise three blocks: a central triple helical block composed of repetitive (Gly-X-Y)$_n$ sequences, and oppositely charged N-terminal and C-terminal blocks containing lysine and aspartate amino acids, respectively. The favorable electrostatic interactions between the terminal blocks have demonstrated to drive the head-to-tail self-assembly of the triblock peptides, leading to the formation of nanofibers. The versatility of triblock collagen mimetic peptides allows for the incorporation of desired functional motifs in the central block, offering exciting prospects for the development of biomaterials with diverse functionalities.

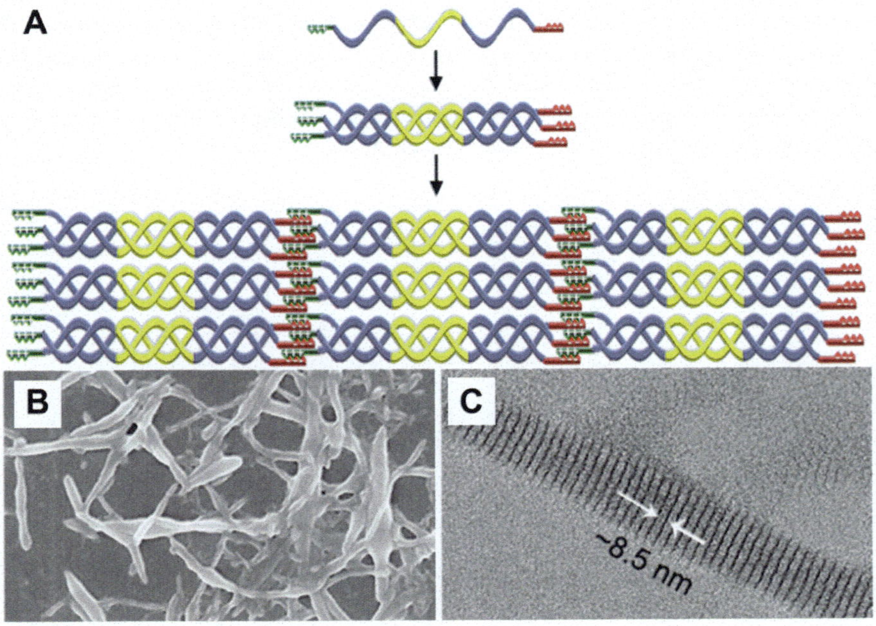

Fig. 12.5 Schematic illustration of the self-assembly of triblock peptides (**A**) and their SEM (**B**) and TEM images (**C**)

12.2.6 Covalent Cross-Linking

Non-covalent interactions have been extensively employed to induce the self-assembly of CMPs into diverse supramolecular structures. Nevertheless, the vulnerability of non-covalent assemblies to environmental fluctuations often results in undesired alterations in their micro/nanostructures, restricting their practical utility. In contrast, covalent self-assembly offers an appealing solution by establishing exceptionally stable peptide assemblies through the creation of irreversible covalent bonds. This approach effectively addresses the constraints associated with non-covalent self-assembly and provides superior structural integrity, rendering it a promising avenue for clinical applications.

Kiick and colleagues developed a collagen peptide that departs from the presence of hydroxyproline but includes cysteine at the C-terminal, reminiscent of type III collagen. The triple helix structure was covalently cross-linked through a simple air oxidation process. The collagen peptide exhibited the characteristic triple helix structure, as confirmed by circular dichroism (CD) spectra, and underwent self-assembly into nanorods and microfiber structures, as evident in TEM images (Fig. 12.6) [34].

Our research group has recently created a novel family of tyrosine-rich triblock peptides capable of covalently self-assembling into highly stable 3D nanofibrous scaffolds. These tyrosine-rich triblock peptides consist of three blocks: a central triple helical block with tyrosine, and two tyrosine-rich N-/C-terminal blocks (designated as Y_N block and Y_C block, respectively). Ruthenium tris-bipyridyl ions ($Ru(II)bpy_3^{2+}$) have been leveraged as catalysts under visible light to initiate the cross-linking of tyrosine residues, facilitating the formation of covalent bonds within the tyrosine-rich triblock peptides. The presence of tyrosine in the triple helical block promotes shoulder-to-shoulder self-assembly, while the terminal tyrosine enhances

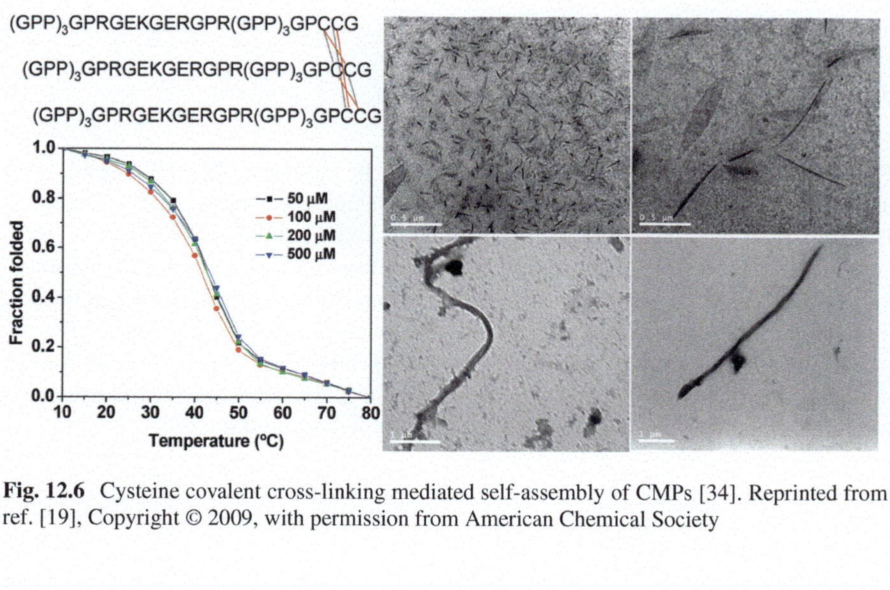

Fig. 12.6 Cysteine covalent cross-linking mediated self-assembly of CMPs [34]. Reprinted from ref. [19], Copyright © 2009, with permission from American Chemical Society

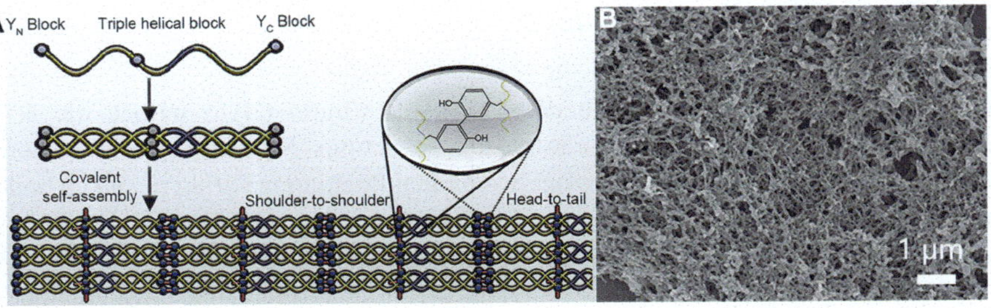

Fig. 12.7 Schematic illustration of the self-assembly of tyrosine-rich triblock peptides (**A**) and the SEM image (**B**)

head-to-tail assembly, synergistically driving the formation of a fibrous network by the tyrosine-rich triblock peptides (Fig. 12.7).

12.3 Self-assembled Heterotrimer Peptides

Significant progress has been made in developing heterotrimeric collagen mimetic peptides (CMPs) consisting of two distinct α chains (AAB) or three different α chains (ABC). However, achieving successful self-assembly remains challenging. Recent investigations have explored innovative strategies, including covalent cross-linking, electrostatic interactions, and Ln^{3+}-ligand interactions, to promote the self-assembly of heterotrimeric peptides into supramolecular structures. Nevertheless, further advancements are needed to fully harness the potential of heterotrimeric peptides in the field of supramolecular self-assembly.

12.3.1 Covalent Cross-Linking

Raines et al. made a groundbreaking discovery by revealing the self-assembly of an ABA-type heterotrimeric peptide through covalent cross-linking of Cys residues in each strand. The incorporation of disulfide linkages resulted in the generation of equivalent numbers of offset triplet triads $(Pro\text{-}Yaa\text{-}Gly)_5$, leading to the presence of overhanging sticky-ends on each strand of the triple helix. This unique design allowed the peptide to exhibit remarkable self-assembly capabilities, forming fibers of varying lengths ranging from 30 nm to over 400 nm [35]. Notably, these synthetic collagen fibers exceeded the length of their natural counterparts, which typically measure around 300 nm, marking a significant milestone in the field.

12.3.2 Electrostatic Interactions

Nanda and colleagues devised novel self-assembled heterotrimeric peptides that leverages electrostatic interactions between oppositely charged amino acids (Fig. 12.8) [36]. Four peptides (cpB-RE, cpC-RE, cpB-KD, and cpC-KD) were synthesized, and only the combinations of cpB-RE with cpC-RE and cpB-RE with cpC-KD resulted in a triple helical structure with a melting temperature (T_m) below 20 °C. TEM analysis revealed that the 1:2 mixture of cpB-RE and cpC-RE possessed the intriguing capability to undergo additional assembly into a fibrous hydrogel. This distinctive behavior can be attributed to the polyvalent guanidyl group of Arg, which promotes both intra- and intermolecular interactions, thereby facilitating an efficient self-assembly process.

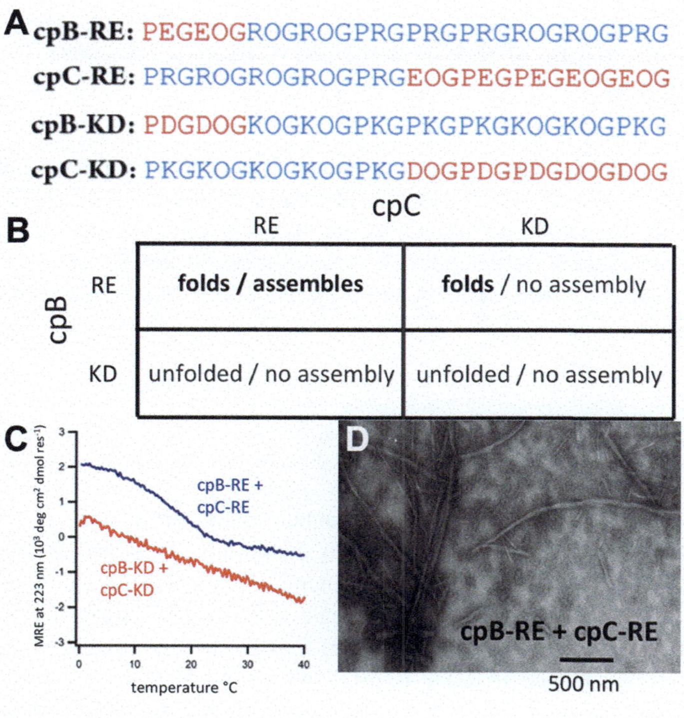

Fig. 12.8 Self-assembly of heterotrimer peptides triggered by electrostatic interactions [36]. Reprinted from ref. [36], Copyright © 2016, with permission from American Chemical Society

12.3.3 *Ln^{3+}-Ligand Interaction*

Our research group has recently reported the Ln^{3+}-mediated self-assembly of a heterotrimeric peptide composed of two A chains (DD(GPRGPO)$_5$DD) and one B chain (DD(GPO)$_{10}$DD). CD and fluorescence analyses demonstrated that despite the electrical repulsion between aspartic acids, the peptides efficiently formed an ABB heterotrimer. Notably, the inclusion of aspartic acids at both the N- and C-termini of the peptides acted as Ln^{3+}-binding ligands, facilitating the self-assembly of the peptides into well-ordered fibers [37]. These findings highlight the pivotal role of lanthanide ions in mediating the self-assembly of heterotrimeric peptides, providing valuable insights into their controlled assembly and potential applications in supramolecular materials.

12.4 Applications of Self-assembled Collagen Mimetic Peptides

Self-assembled peptides have gained prominence as collagen mimetic materials in biomedicine due to their notable benefits, including low immunogenicity, absence of viral contamination, and ease of quality control. These versatile materials can adopt various nanostructures, including fibers, vesicles, spheres, and 3D networks, making them well-suited for a wide array of applications such as drug delivery, cell culture, and hemostasis.

12.4.1 *Cargo Delivery*

Self-assembled CMPs offer significant advantages in the encapsulation of biopolymers by eliminating the harsh preparation conditions associated with synthetic polymer-based microspheres, which can potentially deactivate sensitive biological cargo. Chmielewski and co-workers investigated the encapsulation and delivery of green-fluorescent labeled dextrans (GFLD) using self-assembled CMPs, namely Hbyp3 (with the sequence (POG)$_2$P-Byp-GPOGP-Byp-GPOGP-Byp-G(POG)$_2$) and Hbyp3-11 (with the sequence (POG)$_3$P-Byp-GPOGP-Byp-GPOGP-Byp-G(POG)$_3$) [38]. Upon the addition of Fe^{2+} ions, both peptides underwent self-assembly, resulting in the formation of microcages that effectively encapsulated GFLD.

Confocal microscopy images revealed that the interiors of the microcages were predominantly occupied by the green-fluorescent dextrans, while the collagen peptide cage shells exhibited red fluorescence (Fig. 12.9). Importantly, the release of the encapsulated cargoes from the cages was found to be dependent on the stability of the triple helical peptides, with longer half-lives observed for more stable triple helices

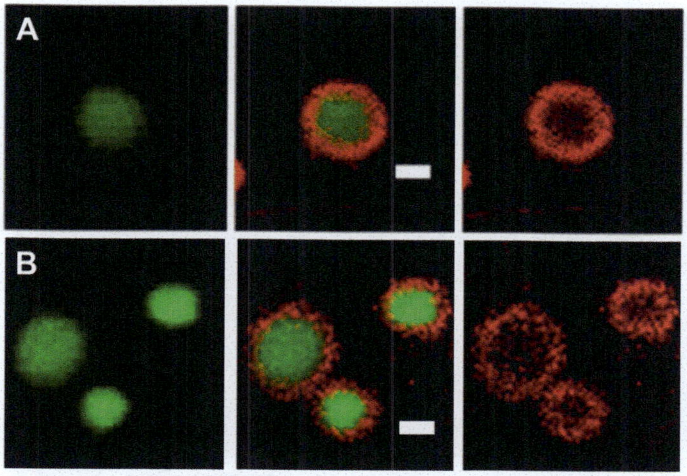

Fig. 12.9 Representative confocal microscopy images for **A** Hbyp3 with 3K MW dextrans and **B** Hbyp3-11 with 40K MW dextrans. Encapsulation of fluorescein-labeled dextrans within the interior of the cages (left), with the exteriors stained with Congo Red (right) and the overlap of the green and red channels (center) (scale bars 1 μm) [38]. Reprinted from ref. [38], Copyright © 2015, with permission from American Chemical Society

[38]. This approach offers an innovative avenue for the development of peptide-based carriers capable of efficiently delivering a wide range of cargoes, including drugs, proteins, and nucleic acids, with promising potential for future biomedical applications.

12.4.2 Cell Culture

The advancement of biocompatible 3D scaffolds capable of encapsulating and nurturing cells is pivotal for progress in tissue engineering and regenerative medicine. Chmielewski and colleagues introduced a metal ion-triggered self-assembling peptide, Nhbipy, and evaluated its capacity for cell encapsulation and cultivation (Fig. 12.10A, B) [26]. Fluorescence microscopy and cryo-SEM imaging provided conclusive evidence of the successful encapsulation of cells within the fibrous network of the collagen peptide-based scaffold, affirming their capacity to proliferate within the scaffold. Notably, the gentle treatment with EDTA facilitated the release of encapsulated cells, which maintained their viability and continued to grow.

Our research group recently reported a series of collagen mimetic peptides with negatively charged amino acids at the middle and N- and C-terminals, showcasing delicate self-assembly into nanofibrous scaffolds upon the addition of lanthanide ions. The incorporation of the integrin-binding motif GFOGER endowed these peptides with exceptional cellular activity. Cell experiments demonstrated that the

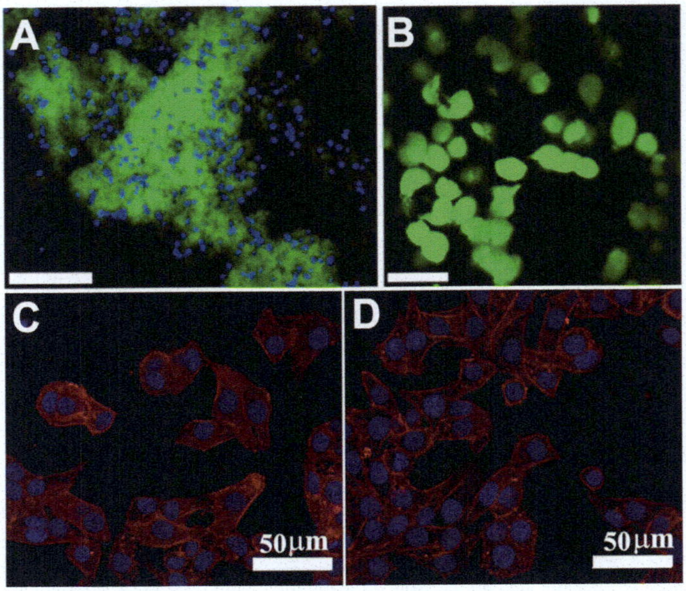

Fig. 12.10 Metal ion mediated self-assembled CMPs for drug delivery: fluorescence microscopy and cryo-SEM images of self-assembled peptides triggered by divalent metal ion (**A**, **B**) [26] (Reprinted from ref. [26], Copyright © 2009, with permission from WILEY–VCH Verlag GmbH & Co. KGaA, Weinheim), and lanthanide ions (**C**, **D**) [39] (Open Access)

self-assembled peptide-based material exhibited cell adhesion capabilities comparable to natural collagen, facilitating the spreading of HeLa cells and effectively mimicking collagen-cell interactions (Fig. 12.10C, D) [39]. These findings highlight the versatility and potential of self-assembled CMPs as a promising platform for cell culture and tissue engineering applications.

12.4.3 Hemostat

The complex interplay between bleeding and clotting processes in traumatic and surgical situations presents a significant challenge to clinicians, highlighting the paramount importance of developing effective hemostatic materials. Due to its inherent thrombogenic properties, collagen has emerged as a promising candidate for hemostatic applications. Hartgerink and colleagues pioneered the utilization of a collagen peptide called KOD, which demonstrated self-assembly into hydrogel and served as a synthetic biomimetic matrix for assessing thrombosis [40]. Their research findings indicated that the self-assembled fibers closely resembled the extracellular matrix, facilitating platelet adhesion, activation, and clotting of plasma and blood in a manner similar to natural collagen. Peptide self-assembly holds promise for creating synthetic hemostatic agents that mimic natural collagen, offering exciting prospects in thrombosis research and hemostatic material development.

References

1. Miller EJ, Gay S (1987) The collagens: an overview and update. Methods Enzymol 144:3–41
2. Shoulders MD, Raines RT (2009) Collagen structure and stability. Annu Rev Biochem 78:929–958
3. Ricard-Blum S (2011) The collagen family. Cold Spring Harb Perspect Biol 3:a004978
4. Sorushanova A, Delgado LM, Wu ZN et al (2019) The collagen suprafamily: from biosynthesis to advanced biomaterial development. Adv Mater 31:e1801651
5. Peter F (2008) Collagen: structure and mechanic. Springer, Germany
6. Kular JK, Basu S, Sharma RI (2014) The extracellular matrix: structure, composition, age-related differences, tools for analysis and applications for tissue engineering. J Tissue Eng 5:2041731414557112
7. Zeltz C, Gullberg D (2016) The integrin-collagen connection—a glue for tissue repair? J Cell Sci 129:653–664
8. Casal JI, Bartolome RA (2018) RGD cadherins and alpha2beta1 integrin in cancer metastasis: a dangerous liaison. Biochim Biophys Acta Rev Cancer 1869:321–332
9. Jokinen J, Dadu E, Nykvist P et al (2004) Integrin-mediated cell adhesion to type I collagen fibrils. J Biol Chem 279:31956–31963
10. Parenteau-Bareil R, Gauvin R, Berthod F (2010) Collagen-based biomaterials for tissue engineering applications. Materials 3(3):1863–1887
11. Li Y, Liu Y, Li R et al (2021) Collagen-based biomaterials for bone tissue engineering. Mater Des 210:110049
12. Copes F, Pien N, Van Vlierberghe S et al (2019) Collagen-based tissue engineering strategies for vascular medicine. Front Bioeng Biotechnol 7:166
13. Fan L, Ren Y, Emmert S et al (2023) The use of collagen-based materials in bone tissue engineering. Int J Mol Sci 24(4):3744
14. Blackstone BN, Gallentine SC, Powell HM (2021) Collagen-based electrospun materials for tissue engineering: a systematic review. Bioengineering 8(3):39
15. Boccafoschi F, Habermehl J, Vesentini S et al (2005) Biological performances of collagen-based scaffolds for vascular tissue engineering. Biomaterials 26(35):7410–7417
16. Luo JN, Tong YW (2011) Self-assembly of collagen-mimetic peptide amphiphiles into biofunctional nanofiber. ACS Nano 5:7739–7747
17. Yao LY, He MM, Li DF et al (2018) Terminal aspartic acids promote the self-assembly of collagen mimetic peptides into nanospheres. Rsc Adv 8:2404–2409
18. Yao LY, He MM, Li DF et al (2018) Self-assembling bolaamphiphile-like collagen mimetic peptides. New J Chem 42:7439–7444
19. Cejas MA, Kinney WA, Chen C et al (2007) Collagen-related peptides: self-assembly of short, single strands into a functional biomaterial of micrometer scale. J Am Chem Soc 129:2202–2203
20. Cejas MA, Kinney WA, Chen CL et al (2008) Thrombogenic collagen-mimetic peptides: self-assembly of triple helix-based fibrils driven by hydrophobic interactions. P Natl Acad Sci USA 105:8513–8518
21. Pires MM, Chmielewski J (2009) Self-assembly of collagen peptides into microflorettes via metal coordination. J Am Chem Soc 131:2706–2712
22. Pires MM, Przybyla DE, Perez CMR et al (2011) Metal-mediated tandem coassembly of collagen peptides into banded microstructures. J Am Chem Soc 133:14469–14471
23. Przybyla DE, Chmielewski J (2008) Metal-triggered radial self-assembly of collagen peptide fibers. J Am Chem Soc 130:12610–+

24. Przybyla DE, Chmielewski J (2010) Metal-triggered collagen peptide disk formation. J Am Chem Soc 132:7866–+
25. Przybyla DE, Perez CMR, Gleaton J et al (2013) Hierarchical assembly of collagen peptide triple helices into curved disks and metal ion-promoted hollow spheres. J Am Chem Soc 135:3418–3422
26. Pires MM, Przybyla DE, Chmielewski J (2009) A metal-collagen peptide framework for three-dimensional cell culture. Angew Chem Int Edit 48:7813–7817
27. He MM, Wang L, Wu J et al (2016) Ln(3+)-mediated self-assembly of a collagen peptide into luminescent banded helical nanoropes. Chem-Eur J 22:1914–1917
28. Yao LY, Hu Y, Liu Z et al (2019) Luminescent lanthanide-collagen peptide framework for pH-controlled drug delivery. Mol Pharmaceut 16:846–855
29. Yao LY, Liu Z, Yu JY et al (2019) Morphology of osteogenesis imperfecta collagen mimetic peptide assemblies correlates with the identity of glycine-substituting residue. ChemBioChem 20:3013–3019
30. Rele S, Song YH, Apkarian RP et al (2007) D-periodic collagen-mimetic microfibers. J Am Chem Soc 129:14780–14787
31. Jiang T, Xu C, Liu Y et al (2014) Structurally defined nanoscale sheets from self-assembly of collagen-mimetic peptides. J Am Chem Soc 136:4300–4308
32. O'Leary LER, Fallas JA, Bakota EL et al (2011) Multi-hierarchical self-assembly of a collagen mimetic peptide from triple helix to nanofibre and hydrogel. Nat Chem 3:821–828
33. Sarkar B, O'Leary LER, Hartgerink JD (2014) Self-assembly of fiber-forming collagen mimetic peptides controlled by triple-helical nucleation. J Am Chem Soc 136:14417–14424
34. Krishna OD, Kiick KL (2009) Supramolecular assembly of electrostatically stabilized, hydroxyproline-lacking collagen-mimetic peptides. Biomacromol 10:2626–2631
35. Kotch FW, Raines RT (2006) Self-assembly of synthetic collagen triple helices. P Natl Acad Sci USA 103:3028–3033
36. Parmar AS, James JK, Grisham DR et al (2016) Dissecting electrostatic contributions to folding and self-assembly using designed multicomponent peptide systems. J Am Chem Soc 138:4362–4367
37. Sun XX, Li WH, Yu JY et al (2020) Ln(3+)-triggered self-assembly of a heterotrimer collagen mimetic peptide into luminescent nanofibers. Chem Commun 56:15141–15144
38. Gleaton J, Chmielewski J (2015) Thermally controlled collagen peptide cages for biopolymer delivery. Acs Biomater Sci Eng 1:1002–1008
39. Sun XX, He MM, Wang L et al (2019) Luminescent biofunctional collagen mimetic nanofibers. ACS Omega 4:16270–16279
40. Kumar VA, Taylor NL, Jalan AA et al (2014) A nanostructured synthetic collagen mimetic for hemostasis. Biomacromol 15:1484–1490

Chapter 13
Future Developments

Collagen mimetic peptides (CMPs) have revolutionized our understanding of collagen's structure and function, significantly contributing to insights into collagen-related diseases. CMP research has progressed from simply replicating collagen's triple helical structure to exploring its wider biological roles. Researchers have employed biophysical techniques such as Circular Dichroism, X-ray Crystallography, Nuclear Magnetic Resonance, Fluorescence, and Raman Spectroscopy to delve into the structural complexities of CMPs. These peptides, initially focused on forming stable triple helices, have shown the ability to self-assemble into complex structures, leading to the creation of biomaterials with collagen-like properties. However, there remains a wealth of untapped potential and opportunities in CMP research.

The field is undergoing a transformation, driven by advances in molecular biology and peptide synthesis. These developments are crucial for producing CMPs that more accurately mimic the diverse types of natural collagen. Traditional CMPs, often limited to about 30 amino acids, have struggled to fully replicate the complex structure and multifunctionality of native collagen. In response, there is a shift towards developing longer and more complex peptide chains, both in homologous and heterotrimeric forms, to deepen our understanding of collagen's roles. Moreover, integrating unique functional motifs into CMPs is gaining focus, enhancing their interactions with various collagen receptors and broadening their biomedical applications.

Advances in biophysical characterization techniques, such as Ultra-High Field NMR spectroscopy, Cryogenic Electron Microscopy (Cryo-EM), Computational Modeling, and Artificial Intelligence (AI), are reshaping CMP research. These sophisticated tools are expected to provide deeper insights into CMPs' structural and functional dynamics.

In the medical field, CMPs are becoming increasingly important, especially in tissue engineering and regenerative medicine, alongside advancements in 3D printing technology. Despite the benefits of biologically derived polymers like modified

© Springer Nature Singapore Pte Ltd. 2024

J. Xiao, *Collagen Mimetic Peptides and Their Biophysical Characterization*,
Springer Series in Biomaterials Science and Engineering 20,
https://doi.org/10.1007/978-981-10-1941-8_13

gelatin and hyaluronic acid, challenges such as variable quality and limited mechanical strength persist. Developing self-assembled CMPs with enhanced mechanical robustness and biocompatibility could revolutionize regenerative medicine and wound healing.

CMPs also hold immense potential in disease pathogenesis, particularly in conditions like cancer, arthritis, and fibrotic diseases. The ability to target pathological collagen specifically could transform disease diagnosis and treatment. However, CMPs currently face limitations in differentiating between various pathological collagen subtypes, an important consideration in complex diseases. Developing methods for precise identification of specific collagen subtypes is vital for disease research and precision treatment.

Pathological collagen's role as a biomarker is promising, but its widespread presence in the body poses challenges. Developing Tissue-Specific Pathological Collagen-Targeting Peptides (PCTPs) could improve disease detection accuracy and sensitivity. In tumor diagnostics, NIR dye-labeled PCTPs have shown promise, and research is progressing towards isotopically functionalized PCTPs for advanced imaging techniques like MRI, CT, and PET imaging. Additionally, PCTPs' affinity for denatured collagen at tumor sites presents opportunities for developing sensitive diagnostic tools. The integration of drugs with CMP-based materials to create targeted prodrugs is an emerging strategy in drug delivery.

The future success of CMPs depends on interdisciplinary collaboration, integrating knowledge from biochemistry, molecular biology, materials science, and clinical research. Tailoring CMPs for personalized medicine is promising, especially in conditions involving collagen dysfunction and tissue engineering. Considering the global burden of collagen-related diseases, advancements in CMP technology could significantly improve global health, offering new treatment options and enhancing patient quality of life.

SPRINGER NATURE

GPSR Compliance

The European Union's (EU) General Product Safety Regulation (GPSR) is a set of rules that requires consumer products to be safe and our obligations to ensure this.

If you have any concerns about our products, you can contact us on ProductSafety@springernature.com

In case Publisher is established outside the EU, the EU authorized representative is:

Springer Nature Customer Service Center GmbH
Europaplatz 3
69115 Heidelberg, Germany

The manufacturer's authorised representative in the EU is Springer Nature Customer Service Centre GmbH, Europaplatz 3, 69115 Heidelberg, Germany. If you have any concerns regarding our products, please contact ProductSafety@springernature.com

Printed and bound by CPI Group (UK) Ltd, Croydon, CR0 4YY
30/09/2026
02225298-0001